TEXTS AND READINGS IN PHYSICAL SCIENCES - 11

The Physics of Disordered Systems

Texts and Readings in Physical Sciences

Managing Editors
H. S. Mani, Chennai Mathematical Institute, Chennai.
hsmani@gmail.com

Ram Ramaswamy, Vice Chancellor, University of Hyderabad, Hyderabad.
r.ramaswamy@gmail.com

Editors

Kedar Damle (TIFR, Mumbai) kedar@tifr.res.in
Debashis Ghoshal (JNU, New Delhi) dghoshal@mail.jnu.ac.in
Rajaram Nityananda (NCRA, Pune) rajaram@ncra.tifr.res.in
Gautam Menon (IMSc, Chennai) menon@imsc.res.in
Tarun Souradeep (IUCAA, Pune) tarun@iucaa.ernet.in

Volumes published so far

1. Field Theories in Condensed Matter Physics, *Sumathi Rao* (Ed.)
2. Numerical Methods for Scientists and Engineers (2/E), *H. M. Antia*
3. Lectures on Quantum Mechanics (2/E), *Ashok Das*
4. Lectures on Electromagnetism, *Ashok Das*
5. Current Perspectives in High Energy Physics, *Debashis Ghoshal* (Ed.)
6. Linear Algebra and Group Theory for Physicists (2/E), *K. N. Srinivasa Rao*
7. Nonlinear Dynamics: Near and Far from Equilibrium, *J. K. Bhattacharjee and S. Bhattacharyya*
8. Spacetime, Geometry and Gravitation, *Pankaj Sharan*
9. Lectures on Advanced Mathematical Methods for Physicists, *Sunil Mukhi and N. Mukunda*
10. Computational Statistical Physics, *Sitangshu Bikas Santra and Purusattam Ray* (Eds.)

The Physics of Disordered Systems

Edited By
Gautam I. Menon
and
Purusattam Ray

Published by

Hindustan Book Agency (India)
P 19 Green Park Extension
New Delhi 110 016
India

email: info@hindbook.com
www.hindbook.com

Copyright © 2012, Hindustan Book Agency (India)

No part of the material protected by this copyright notice may be reproduced or utilized in any form or by any means, electronic or mechanical, including photocopying, recording or by any information storage and retrieval system, without written permission from the copyright owner, who has also the sole right to grant licences for translation into other languages and publication thereof.

All export rights for this edition vest exclusively with Hindustan Book Agency (India). Unauthorized export is a violation of Copyright Law and is subject to legal action.

ISBN 978-93-80250-32-8

Texts and Readings in the Physical Sciences

The Texts and Readings in the Physical Sciences (TRiPS) series of books aims to provide a forum for physical scientists to describe their fields of research interest. Each book is intended to cover a subject in a detailed manner and from a personal viewpoint, so to give students in the field a unique exposure to the subject that is at once accessible, pedagogic, and contemporary. The monographs and texts that have appeared so far, as well as the volumes that we plan to bring out in the coming years, cover a range of areas at the core as well as at the frontiers of physics.

In addition to texts on a specific topic, the TRiPS series includes lecture notes from a thematic School or Workshop and topical volumes of contributed articles focussing on a currently active area of research. Through these various forms of exposition, we hope that the series will be valuable both for the beginning graduate student as well as the experienced researcher.

H.S. Mani
Chennai

R. Ramaswamy
Hyderabad

In memory of

Rahul Basu

physicist, editor, and friend

Contents

Preface   ix

List of Contributors   xi

1 Introduction to the Physics of Disordered Systems   1
Gautam I. Menon and Purusattam Ray
  1.1 Thermodynamics and the Link to Statistical Mechanics 1
  1.2 Statistical Mechanics: A Summary of Important Results ... 5
  1.3 Mean Field Theory and Phase Transitions in Pure Systems . 9
  1.4 Landau-Ginzburg Theory of Phase Transitions 13
  1.5 The Percolation Problem and Geometric Phase Transitions.. 20
  1.6 Disordered Spin Systems 23
  1.7 Spin Glass Basics: Experimental Phenomenology 30
  1.8 Spin Glass Physics: Towards a Model 32
  1.9 The Sherrington Kirkpatrick Model 34
  1.10 Physics of the Spin Glass State 41
  1.11 Modern Developments 43
  1.12 K-SAT and the Spin Glass Problem 45
  References 47

2 Phase Transitions in Disordered Quantum Systems: Transverse Ising Models   49
Bikas K. Chakrabarti and Arnab Das
  2.1 Transverse Ising Model 51
  2.2 Mean Field Theory 51
  2.3 BCS Theory of Superconductivity 55
  2.4 Real Space Renormalization for the Transverse Ising Chain . 59
  2.5 Equivalence of d-dimensional Quantum Systems and $(d+1)$-dimensional Classical Systems: Suzuki-Trotter Formalism ... 63
  2.6 Transverse Ising Spin Glasses 66
  2.7 Quantum Annealing 77
  2.8 Summary and Discussions 81
  References 83

3 The Physics of Structural Glasses — 85
Srikanth Sastry

- 3.1 Introduction — 85
- 3.2 Phenomenology of Glass Formation — 87
- 3.3 Computer Simulations — 94
- 3.4 Theoretical Approaches — 97
- References — 114

4 Dilute Magnets — 121
Deepak Kumar

- 4.1 Introduction — 121
- 4.2 Percolation Processes — 122
- 4.3 Dilute Magnets: Scaling Theory — 125
- 4.4 Potts Model and Percolation — 129
- 4.5 Diluted Ising Model — 132
- 4.6 Neutron Scattering Results and a Simple Model of Correlations — 135
- References — 139

5 Domains and Interfaces in Random Fields — 141
Prabodh Shukla

- 5.1 Introduction — 141
- 5.2 Domains in Quenched Random Fields — 142
- 5.3 Relevance to Experiments — 145
- 5.4 Roughness of Interfaces — 147
- 5.5 Absence of Order in 2d RFIM — 149
- 5.6 A Toy Problem — 152
- 5.7 Thermal Effects — 154
- References — 158

6 Vortex Glasses — 159
G. Ravikumar

- 6.1 Introduction — 159
- 6.2 Irreversibility in Type II Superconductors — 162
- 6.3 Magnetic Relaxation and Flux Creep — 165
- 6.4 Irreversibility Line — 168
- References — 170

Index — 173

Preface

Why should we care about disordered systems?

First, because virtually *every* physical system of experimental relevance is disordered: Crystals achieve a perfectly periodic arrangement of atoms only in the inaccessible zero temperature limit. Materials which superconduct at relatively high temperatures require their insulating parent compounds to be doped away from chemical stoichiometry. Structural glasses, such as window glass, behave like solids to mechanical perturbations but are amorphous, resembling fluids, in their structure. The semiconductor chips at the heart of virtually all of modern technology originate in the incorporation of controlled amounts of impurities into ultrapure silicon.

Second, while the textbook paradigm of the pure material is often a useful starting point, much of the interesting physics of disordered systems, such as metal-insulator transitions as a consequence of electron localization, the spin and structural glass transitions, and the percolation problem, is to be found in limits where the effects of disorder cannot be studied by perturbing away from a pure system. Understanding disordered systems thus requires the development of new, often radically different intuition, as well as of calculational methods with no counterpart in the study of pure materials.

Third, methodologies inspired by the study of disordered systems have long spurred developments in fields as diverse as theoretical computer science, optimization theory, neuroscience and protein folding. The relationship between satisfiability problems in theoretical computer science and modern replica-based approaches to spin glasses is an interesting example of how results in one field can have important implications for other seemingly disconnected areas.

In what way does the study of disordered systems differ from the study of conventional, even complicated non-disordered ones? Principally because such disorder enters at the microscopic level and arises from distributions of microscopic variables that cannot be accessed directly. This requires that we perform an appropriate average over the variables representing the disorder. In some cases, this averaging resembles the customary averaging employed in the statistical mechanics of pure systems. However, the far more interesting case is when the variables characterizing the disorder behave in a qualitatively different manner from the dynamical variables of the problem. This is the case of *quenched* disorder. The difficulties of averaging over quenched disorder lie at the heart of disordered systems physics.

This book is a collection of pedagogical articles describing the physics of disordered systems. Its original inspiration was an summer school program, sponsored by the DST (India) under the SERC program and organized by both of us. The school targeted graduate students either working in or intending to enter the field of the physics of disordered systems. Several of the lecturers agreed to contribute chapters for a book which would address our current understanding of a variety of disordered systems of relevance both to the experimentalist as well as to the theorist.

The school, and this book, were motivated by the following: The problems faced by a new entrant into the field of disordered systems are formidable. One must first have some intuition for the physics of non-disordered systems to be able to appreciate the alterations introduced by disorder. The field of disordered systems is itself a large and specialized field, with much terminology that is unique to the area. However, we know of no single textbook at the graduate level which a student can use to learn more about areas subsumed under this broad label. Thus, while excellent books about, say, spin glasses exist, such books will typically not discuss the physics of structural glasses. This book is an attempt to fill that gap.

The Department of Science and Technology (DST), India (SERC Schools Program) and the Institute of Mathematical Sciences, Chennai (PRISM project) provided generous support, for which we are very thankful. We are grateful to the lecturers at the school who wrote up detailed notes for their lectures and took the trouble to translate them into a level appropriate to a textbook of this sort. We are especially indebted to Ram Ramaswamy for "keeping the faith" and believing that this book would be out one day, despite all evidence to the contrary. Several colleagues commented on draft versions of the book and its chapters and we thank them for their input.

Gautam I. Menon and Purusattam Ray

The Institute of Mathematical Sciences
Chennai, 2011

List of Contributors

- *Bikas Chakrabarti* (bikask.chakrabarti@saha.ac.in) is at the Saha Institute of Nuclear Physics, Kolkata 700 064

- *Arnab Das* (arnabdas@pks.mpg.de) is at the Max Planck Institute for the Physics of Complex Systems, Nöthnitzer Str. 38, 01187 Dresden, Germany

- *Deepak Kumar* (dkjnu@hotmail.com) is at the School of Physical Sciences, Jawaharlal Nehru University, New Delhi 110 067

- *Gautam I. Menon* (menon@imsc.res.in) is at the Institute of Mathematical Sciences, CIT Campus, Taramani, Chennai 600 113

- *G Ravikumar* (gurazada@apsara.barc.ernet.in) is at the Technical Physics Division, Bhabha Atomic Research Centre, Mumbai 400 085

- *Purusattam Ray* (ray@imsc.res.in) is at the Institute of Mathematical Sciences, CIT Campus, Taramani, Chennai 600 113

- *Srikanth Sastry* (sastry@jncasr.ac.in) is at the Jawaharlal Nehru Centre for Advanced Scientific Research, Jakkur campus, Bangalore 560 064

- *Prabodh Shukla* (shukla@nehu.ac.in) is at the Physics Department, North Eastern Hill University, Shillong 793 022

Chapter 1

Introduction to the Physics of Disordered Systems

Gautam I. Menon and Purusattam Ray

This chapter reviews some basic material required for the study of disordered systems, establishing some of the terminology used in later chapters. It contains a summary of the necessary background in statistical mechanics, including mean-field theory, scaling at continuous phase transitions, and Landau theory. It describes the percolation problem as well as provides informal derivations of some useful general results in disordered systems, such as the Harris criterion, the Imry-Ma argument and Griffiths singularities. Spin glasses are introduced as a prototypical example of disordered systems. The replica approach to the Sherrington-Kirkpatrick model of infinite range Ising spin glasses is briefly described, together with a summary of more recent developments in spin glass physics. The connection between theories of spin glasses and satisfiability problems in theoretical computer science is also outlined.

1.1 Thermodynamics and the Link to Statistical Mechanics

Thermodynamics is grounded in the following assertion: There is a class of macroscopic states of physical systems in which measured properties are independent of preparation history. Such states, if they can be consistently described through the thermodynamic formalism discussed below, are defined to be at **thermal equilibrium**.

The physical properties of systems in thermal equilibrium, at a sufficiently coarse-grained level, are determined by a small number of **macroscopic variables**. Two distinct systems in thermal equilibrium which possess identical values of these variables will behave identically as far as macroscopic measurements are concerned. Such systems can be said to be in the same **macrostate**. A complete set of macroscopic variables describing a monoatomic gas are the number of molecules N, the volume V and the internal energy E of the gas. These are **extensive variables** *i.e.* for a composite system made up of two subsystems, each having its own value of E, V and N, the values of E, V and N characterizing the composite system are obtained by adding their values for the subsystems.

The internal energy E of a thermodynamic system can be changed in several ways. One could modify, for example, a **mechanical variable**, such as the volume V of the system. Alternatively, one could alter a **chemical variable**, such as the number of molecules N of a specific type. The central contribution of thermodynamics is to recognize that there is another way of changing E which does not involve changes in either mechanical or chemical variables. This involves the transfer of heat to or from the system. Operationally, heat may be transferred by placing the system in contact with another macroscopic system at a different temperature, in which case energy is transferred from the body at the higher temperature to the one at a lower temperature.[1] The transfer of heat to or from a system alters its internal energy without changing N and V. Understanding the physical consequences of heat transfer requires the introduction of another extensive variable, the **entropy**.[2]

Thermodynamics postulates that a single function of the extensive variables suffices to fully characterize the thermodynamic behaviour of the system. Such a function, termed a **fundamental relation**, can be written as $S = S(E, N, V)$, defining a quantity called the entropy S as a function of E, V and N. Postulating the existence of S, coupled together with a few additional, reasonable assumptions concerning its monotonicity and analyticity, reproduces all the classical results of thermodynamics. In textbooks, this discussion is typically presented in terms of Carnot engines and cyclic thermodynamic processes which take the engine through a sequence of thermodynamic states. However, the approach of *postulating* the existence of an entropy *a priori* is certainly more elegant.

Thermodynamics is concerned only with macroscopic quantities. However, macroscopic behaviour is determined by microscopic configurations and their relative weights. The entropy S is the single central quantity which connects the description of the system at the level of its microstates to the macrostate specification. Adding heat to the system, keeping other macroscopic variables of the system such as N and V constant, changes the distribution of the system across its microstates, thereby altering the entropy. It is natural to assume that S is a monotonically increasing function of the internal energy E. (Given that S is a monotonic function of E, the relationship can be reversed to provide

---

[1] The existence of such a quantity is a fundamental postulate of thermodynamics, the zeroth law, which identifies the equality of temperatures as a condition for two bodies in thermal contact to be in thermal equilibrium.

[2] The fact that the internal energy is conserved is the first law of thermodynamics, reflecting the fact that the chemical and mechanical work done on the system, when added to the heat transferred to the system, must exactly equal the change in the systems internal energy.

$E = E(S, N, V)$, a form which is often more convenient.) For a composite system, made of two independent systems with extensive variables N_1, V_1 and E_1 and N_2, V_2 and E_2, and thus the fundamental relations $S_1(E_1, N_1, V_1)$ and $S_2(E_2, N_2, V_2)$, the entropy S is additive *i.e.*

$$S(E = E_1 + E_2, N = N_1 + N_2, V = V_1 + V_2 = S_1 + S_2) \quad (1.1)$$
$$= S(E_1, V_1, N_1) + S(E_2, V_2, N_2)$$

Provided the subsystems composing the composite system do not exchange energy, particles or volume, these quantities are separately conserved for the subsystems.

Having defined the entropy as an additional macroscopic variable required to describe thermodynamic systems, its utility emerges from the following assertion: If the two subsystems are allowed to exchange energy, ensuring that only $E = E_1 + E_2$ is conserved and not E_1 and E_2 separately for each subsystem, then the system seeks a final thermodynamic state such that $S = S(E, V, N)$ is maximized over all ways of partitioning E over the subsystems. (If the subsystems exchange particles or volume, a similar statement applies to $N = N_1 + N_2$ and to $V = V_1 + V_2$.)

We can think of relaxing the constraint that the two subsystems do not exchange energy, volume or particle number as relaxing an internal constraint of the composite system. The second law of thermodynamics can then be phrased in the following way:

> Removing an internal constraint results in a final system entropy which cannot be lower than the initial entropy of the constrained system

An **extremum principle**, the maximization of entropy, thus controls the equilibrium behaviour of a thermodynamic system. This is just the law of entropy increase, described in conventional textbook formulations, of the **Second Law of thermodynamics**.

From this, all of thermodynamics can be shown to follow, including the definitions of temperature T, chemical potential μ and pressure P. These definitions are

$$T = \left(\frac{\partial E}{\partial S}\right)_{N,V}, \quad P = -\left(\frac{\partial E}{\partial V}\right)_{S,N}, \quad \mu = \left(\frac{\partial E}{\partial N}\right)_{S,V}. \quad (1.2)$$

Such a definition ensures that two systems which can exchange energy are in thermal equilibrium only if their temperatures are equal. Similarly, mechanical equilibrium between two systems which can exchange matter (volume) requires that the chemical potentials (pressures) of both subsystems be equal.

In calculations or in experiments, it is often easier to constrain the temperature T rather than E. This can be done by allowing the system to exchange energy with a **thermal bath** at temperature T. Such a bath can be realized in terms of a system with a very large number of degrees of freedom, whose temperature is unchanged if an arbitrary but finite amount of energy is added to or removed from it. Defining an appropriate extremum principle if T is fixed, can be accomplished using a Legendre transform. Since T and S are conjugate variables, the Legendre transform $F(T, V, N)$ is defined through

$$F(T, V, N) = E(S, V, N) - TS. \quad (1.3)$$

The Legendre transform F of E, the **Helmholtz free energy**, provides a new fundamental relation in terms of the quantities T, V and N. Like the entropy, it provides a complete thermodynamic description of the system. It also obeys a similar extremum principle – the Helmholtz free energy is minimized in thermal equilibrium. Other potentials follow by eliminating other conjugated quantities e.g. the number of particles N in favour of the chemical potential or the volume V in favour of the pressure P.

The concept of entropy is central to thermodynamics and changes in the entropy represent the effects of the transfer of energy in the form of heat to microscopic degrees of freedom. In a theory which centers on these microscopic degrees of freedom (statistical mechanics), the entropy must have a suitable microscopic definition. The crucial connection between microscopic and macroscopic behaviour was established by Boltzmann, who first wrote down the fundamental formula

$$S = k_B \ln \Omega(E, V, N), \quad (1.4)$$

with $\Omega(E, V, N)$ defined as the number of microstates accessible to a system once a fixed E, V and N are specified. The quantity k_B is Boltzmann's constant, required for consistency with the units in which thermodynamic measurements are conventionally made. The numerical value of k_B is 1.38×10^{-23} J/K.

This defines, for statistical mechanics, the fundamental relation of the system, in the **microcanonical ensemble**. In calculations within this statistical ensemble, all configurations with the same energy, number of particles and volume are assigned equal weight. Given this fundamental relation, other thermodynamic potentials can be easily derived, for example corresponding to fixing a temperature as opposed to fixing the system energy. Since all ensembles carry identical information and one can move from ensemble to ensemble using Legendre transformations, results for thermodynamic quantities must be the same in different ensembles in the thermodynamic limit. This is a basic consistency requirement called the **equivalence of ensembles**. This mandates that the most probable value and the mean value of thermodynamically measurable quantities should coincide in the limit of large systems. If they do not coincide at isolated points, this is a signal that fluctuations dominate the behaviour at those points. Typically, this is a concern only when a quantity which is fixed in one ensemble has divergent fluctuations in another.

While the broad formalism of classical thermodynamics carries over to the study of disordered systems, such systems do have some unique features. For example, disordered systems often possess a large number of states infinitesimally close to (or possibly even degenerate with, depending on the model) the ground state, the state which represents the absolute minimum of the energy. While this is also a possibility for non-disordered models, these different states in disordered systems need not be simply related by symmetry. Thus, appropriately defined coarse-grained free energies in disordered systems typically have more complex structure than their pure system counterparts.

Such free energies can possess multiple low-lying minima (metastable states) in the space of configurations, separated by large barriers. A metastable state, in the context of the spin models we discuss later, is one whose energy is unchanged or only changes to order unity, when a finite number of spins are flipped. The degeneracy of these states can, in some cases, be extensive in the size of the system. Slow relaxation out of such metastable states often dominates the dynamics, ensuring that the system does not equilibrate over experimental time scales, thus obscuring the relationship to the underlying equilibrium system.

1.2 Statistical Mechanics: A Summary of Important Results

A physical system is characterized by listing the states available to the system, as well as by specifying a dynamics which takes the system between these states. For example, for a system of N interacting classical particles in three dimensions, the state of the system at each instant is specified through the momenta and positions of each of the N particles. The space of possible states is simply the space of all possible momenta and positions of the N particles. The dynamical evolution of a classical system initially in one of these states is provided by Newton's equations. Integrating these equations forward in time specifies a trajectory in phase-space. Any thermodynamic quantity can be obtained at any instant as a function of the $6N$ (in three dimensions) phase space variables. As the system evolves in time, a thermodynamic average can be constructed as an average over such measurements evaluated along the trajectory.

For systems described by conventional thermodynamics, one assumes **ergodicity**. This corresponds to the statement that time averages obtained by following the system along its phase space trajectories (and thus a property of the dynamics of the system), can be replaced by a statistical average over the probabilities of finding the system in its different states. If one creates a large number of copies of the system in its different states, with each such state represented in proportion to the probability of finding that state in thermal equilibrium, this defines what is called an **ensemble**. The average of thermodynamic quantities in such an ensemble is called an ensemble average. Thus, the assumption that time-averages equal ensemble averages is the statement of ergodicity. Physically, this is the same as saying that the information in any sufficiently long trajectory is the same, as far as thermodynamics is concerned, as the information in a sequence of independent snapshots of the system, provided there a sufficiently large number of them for probability distributions to be well defined.

For concreteness, consider Ising spin models, possibly the simplest models for interacting systems. These are models for spins placed on a lattice, in which the spins are constrained to take values $\sigma_i = \pm 1$. A model in which N Ising variables σ_i interact pairwise with each other as well as separately with an external field H_{ext} is

$$\mathcal{H} = -J \sum_{\langle ij \rangle}^{N} \sigma_i \sigma_j - H_{ext} \sum_{i=1}^{N} \sigma_i. \qquad (1.5)$$

Interactions between spins are modeled via an exchange coupling J. The notation $\langle ij \rangle$ normally implies a sum over distinct pairs of spins. This sum is often restricted to the case in which the pairs considered are neighbouring sites. The lowest energy state of this model, or its **ground state**, has all Ising spins aligned with the external field i.e. $\{\sigma_i = 1\}$. In the absence of the external field, both the all-up state and the all-down state are degenerate ground states. The Ising model is the simplest example of a solvable model which exhibits non-trivial behaviour as a consequence of interactions. The Ising model exhibits a finite temperature phase transition at zero field in dimensions $d = 2$ and above.

A microstate of this model is a particular spin configuration, for example the configuration:

$$\{01010101100000101010\ldots\}. \qquad (1.6)$$

There are 2^N microstates in this simple model.

Statistical mechanics prescribes rules for connecting averages over microstates to macrostates. From these averages, we can calculate thermodynamic properties such as the specific heat, the average energy or susceptibility. Ensembles in statistical mechanics correspond to specific rules for assigning weights to microstates, depending on the constraints placed on the system.

The microcanonical ensemble, discussed in the previous section, refers to the case of an isolated system, with fixed energy. An experimentally more accessible system is one in which the system is allowed to exchange energy with the environment, thus coming into thermal equilibrium with it. Thus, the system and the environment (the bath) have the same temperature. In the **canonical ensemble**, intended to describe precisely such systems placed in contact with a large bath at fixed temperature, the rule for generating the ensemble of microstate configurations is that each microstate k of energy E_k is assigned a probability p_k with

$$p_k = \exp(-\beta E_k)/\mathcal{Z}, \qquad (1.7)$$

where $\mathcal{Z}$ is a normalizing factor and $\beta = 1/k_B T$.

1.2. Statistical Mechanics: A Summary of Important Results

Since we know the probability of finding the system in any of its microstates, any macroscopic quantity may be obtained as a sum of its value in each microstate, weighted by this probability. Thus

$$\langle O \rangle = \sum_k O(k) p_k, \qquad (1.8)$$

where the sum ranges over all microstates and $O(k)$ is the value of the observable O in microstate k. We assume that the probabilities are appropriately normalized, so that $\sum_k p_k = 1$.

The calculation of thermodynamic quantities is simplified through the calculation of the **partition function** $\mathcal{Z}$, defined by

$$\mathcal{Z} = \sum_{\{\sigma\}} e^{-\beta E}. \qquad (1.9)$$

where we sum over all values the spins can take, *i.e* over all the microstates of the system. Sometimes, for notational reasons, we replace $\sum_{\{\sigma\}}$ by the trace notation Tr_σ, but the calculational operation is the same in both cases. The quantity $\mathcal{Z}$ is the same as the normalization factor required in the calculation of the normalized p_k but its value far exceeds that of a mere normalization.

The link between the partition function and the thermodynamic potential of relevance here, the Helmholtz free energy (since the system is in contact with a heat bath at temperature T), is

$$F = -k_B T \ln \mathcal{Z}. \qquad (1.10)$$

The free energy $F(T, V, N)$ yields all thermodynamic properties of interest. Other free energies can be calculated simply through appropriate Legendre transformations.

The calculational bottleneck is almost invariably the sum over microstates. The difficulty arises from the large number of microstates to be considered. Even for 100 spins, the number of states is $2^{100} \sim 10^{30}$, which is prohibitively large. In some special cases, this can be circumvented, as for example in the one-dimensional Ising model which can be solved exactly. In two dimensions, this is a technically formidable calculation even in the simplifying limit where H_{ext} is zero, an accomplishment due to the Norwegian physical chemist Lars Onsager in the 1940's. In the presence of a non-zero applied magnetic field and in dimensions greater than two for the zero-field case, it has so far proved impossible to solve the model exactly. The solution of the two-dimensional Ising model in a non-zero external field remains an outstanding problem in statistical mechanics.

1.2.1 Quenched and Annealed Disorder

The formulation of the statistical mechanics of disordered systems requires an additional ingredient, the distinction between **quenched** and **annealed** disorder.

Apart from the dynamical variables (the σ's) themselves, one could imagine that the coupling constant J varied from site to site, i.e. that a J_{ij} depending on the choice of neighbouring sites i and j replaced the uniform J,

$$\mathcal{H} = -\sum_{\langle ij \rangle}^{N} J_{ij} \sigma_i \sigma_j - H_{ext} \sum_{i=1}^{N} \sigma_i. \tag{1.11}$$

The energy thus depends both on the configuration of the spin variables and the configuration of the J_{ij}'s. We could fix the J_{ij}'s at the outset, but such quantities are typically not directly accessible in experiments. On the other hand, it is often feasible and realistic to compute the *distribution* of these coupling constants. That is, given a set of J_{ij}'s, one can compute the probability that such a set is realized, $P(\{J_{ij}\})$. Thus we must compute two averages over probability distributions. The first is the average over spin configurations in say, the canonical ensemble, within a fixed background of J_{ij}'s and the second is the average over the probability distribution of the J_{ij}'s.

If the disorder in the coupling constants enters on the same footing as the fluctuating dynamical variables, the disorder is termed as **annealed**. The variables representing such disorder can be summed over in the standard way when taking the trace in the partition function, in the following way.

$$Z_{annealed} = \int \prod dJ_{ij} P(\{J_{ij}\}) Z(\{J_{ij}\}). \tag{1.12}$$

Technical difficulties aside, this is really the same as computing a partition function involving a larger number of degrees than those arising from the spins alone. (This just expands the ensemble of spin configurations to include the ensemble of disorder configurations arising from having different values of the J_{ij}'s, weighted appropriately.) **Quenched** disorder, on the other hand, refers to quantities which appear in the definition of the partition function but which are *not dynamical* and thus are not summed over. Thus, for quenched disorder in the J_{ij}'s, we can compute the partition function in a fixed background of J_{ij}'s by summing over spin configurations. We need to decide whether this partition function is to be simply averaged over disorder, as is the case for annealed disorder, or whether it is the free energy, proportional to the logarithm of the partition function, which is to be averaged. It will emerge that this second possibility is the correct way of averaging over quenched disordered variables.

Conventional statistical mechanics approaches work in the case of annealed variables. They just introduce more variables into the partition sum. At a basic level, however, the statistical mechanical problem is very similar. In contrast to this, the appearance of quenched variables in the theory introduces qualitatively new physics.

1.3 Mean Field Theory and Phase Transitions in Pure Systems

Phase transitions separate regimes of behaviour in which a suitably defined macroscopic variable of thermodynamic significance, called an **order parameter**, changes from being zero in the disordered phase to a non-zero value in the ordered phase. More generally, phase transitions represent singularities in the free energy which appear in the thermodynamic limit. Order parameters can, however, be subtle. A significant achievement of the Landau-Ginzburg theory of phase transitions in superfluid helium and superconductors was the realization that an order parameter need not be a directly experimentally measurable quantity. Sometimes no appropriate *local* definition of an order parameter exists and an appropriate non-local definition must be sought. The issue of a suitable order parameter for the sorts of disordered systems we will encounter is a delicate one. At least in one celebrated case, that of replica symmetry breaking in the Sherrington-Kirkpatrick Ising spin glass model, the order parameter is a function in the interval $[0, 1]$.

A simple and intuitive approach to understanding several important properties of the model, including the appearance of a phase transition, is that of **mean-field theory**. The rationale for this approach is the following: One can virtually always solve the problem of a *single degree of freedom* in an external field, since there is only one variable here as opposed to 10^{30}. So one strategy would be to attempt to reinterpret the original interacting spin problem in terms of the problem of a single spin in an effective magnetic field. This field is determined self-consistently, a step which lies at the heart of all mean-field approaches.

1.3.1 Single Site Mean-field Theory

Start with the Ising Hamiltonian in d-dimensions, on a hypercubic lattice, with the energy

$$\mathcal{H} = -J \sum_{\langle ij \rangle}^{N} \sigma_i \sigma_j - H_{ext} \sum_{i=1}^{N} \sigma_i. \tag{1.13}$$

Consider a given spin, with z neighbours. It interacts with these spins via the exchange term J. Each spin also interacts separately with an external field H_{ext}. (For simplicity, we will set the external field to zero.) We start with the obvious result

$$\sigma_i = \sigma_i + m - m. \tag{1.14}$$

We assume that

$$\langle \sigma_i \rangle = \langle \sigma_j \rangle = m, \tag{1.15}$$

and that we can neglect fluctuations, thus ignoring terms quadratic in the fluctuation, that is, terms of the form $(\sigma_i - m)(\sigma_j - m)$ Thus

$$\sigma_i\sigma_j = (\sigma_i - m + m)(\sigma_j - m + m) \simeq m^2 + m(\sigma_i - m) \quad (1.16)$$
$$+ m(\sigma_j - m) = m(\sigma_i + \sigma_j) - m^2.$$

Now

$$\mathcal{H} - \frac{JNzm^2}{2} = -J\sum_{\langle ij \rangle}\sigma_i\sigma_j = -\frac{Jzm}{2}\sum_i^N (2\sigma_i) \quad (1.17)$$
$$= -Jzm\sum_i^N \sigma_i = -Jzm\sum_i^N \sigma_i.$$

This then yields

$$Z = \exp\left(-\beta JzNm^2/2\right)(2\cosh(\beta Jzm))^N. \quad (1.18)$$

The magnetization, m is defined by

$$m = \sum_i \langle \sigma_i \rangle / N. \quad (1.19)$$

The average value of the spin σ_i must then self-consistently be the magnetization, yielding

$$m = \tanh(\beta Jzm), \quad (1.20)$$

a self-consistent equation which must be solved for m.

The solution to this equation can be obtained graphically at any value of β. There is no solution except at $m = 0$ for small β (high temperatures). Two solutions $\pm m_0$ appear at sufficiently low temperatures, with $m_0 \to 1$ as $T \to 0$, as shown in Fig. 1.1. Close to the transition between these two different types of behaviour, which occurs at a non-zero and finite value of β, the value of m is small and one can expand

$$\tanh(\beta Jmz) = \beta Jmz + \frac{(\beta Jmz)^3}{3} + \ldots, \quad (1.21)$$

which yields the equation determining m

$$m(1 - \beta Jz) = \frac{(\beta Jmz)^3}{3}. \quad (1.22)$$

Note that $m = 0$ always trivially solves this equation. However, a non-trivial solution is obtained at values of β larger than a critical value β_c, with

$$\beta_c = 1/Jz, \quad (1.23)$$

1.3. Mean Field Theory and Phase Transitions in Pure Systems

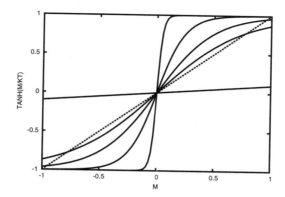

Figure 1.1: Plot of the self-consistent solution to $tanh(\beta M) = M$, obtained by plotting the two curves with different values of β. For small β, i.e. large temperatures, the curves intersect only at $M = 0$, indicating that only one (stable) solution to the self-consistent equation exists. Below a critical temperature, while the solution at $M = 0$ is always present, additional non-trivial solutions at nonzero $\pm M_0$ are obtained, which evolve smoothly out of the $M_0 = 0$ state as β is increased. The states with non-zero magnetization are lower free energy states below the critical point.

implying a critical temperature T_c, given by

$$k_B T_c = Jz. \qquad (1.24)$$

Below this critical temperature, the magnetization evolves smoothly from zero, saturating to unity at $T = 0$. Such a smooth, continuous evolution of the order parameter from zero to non-zero across the transition, characterizes a **continuous phase transition**.

Note that mean-field theory predicts a transition in all dimensions, since dimensionality enters only indirectly in this argument, via the relation between z and d.

1.3.2 Mean-field Theory Via the Bragg-Williams Approximation

We now discuss the infinite range version of the model above, with the Hamiltonian

$$\mathcal{H} = -\frac{J}{2N} \sum_{ij} \sigma_i \sigma_j \qquad (1.25)$$

Here the sum is over all pairs. The factor of N ensures proper extensivity, since the all aligned $T = 0$ state must have an energy which is properly extensive in N.

We have

$$m^2 = \left(\frac{1}{N}\sum_i \sigma_i\right)\left(\frac{1}{N}\sum_j \sigma_j\right) = \left(\frac{1}{N^2}\sum_{ij} \sigma_i \sigma_j\right), \quad (1.26)$$

from which $\sum_{ij} \sigma_i \sigma_j = N^2 m^2$. Then we may rewrite the Hamiltonian as

$$\mathcal{H} = -\frac{NJm^2}{2}. \quad (1.27)$$

We can now integrate over all allowed m with the appropriate weight. To do this, we calculate the number of configurations with a fixed number of up and down spins consistent with a given magnetization. Given $m = (N_{up} - N_{down})/N$ and $N = N_{up} + N_{down}$, we can write the entropy as

$$S = \ln \frac{N!}{N_{up}! N_{down}!}, \quad (1.28)$$

the standard Bragg-Williams form, which is then

$$\begin{aligned} S &= \ln\left(\frac{N!}{(N(1+m)/2)!(N(1-m)/2!)}\right) \\ &= N \times \left[\ln 2 - \frac{1}{2}(1+m)\ln(1+m) - \frac{1}{2}(1-m)\ln(1-m)\right]. \end{aligned} \quad (1.29)$$

Thus we may write the free energy per spin as

$$f = \frac{F}{N} = -\frac{1}{2}Jm^2 + \frac{T}{2}\left[(1+m)\ln(1+m) + (1-m)\ln(1-m)\right] - T\ln 2. \quad (1.30)$$

The equation of state follows from $\partial f/\partial m = 0$ after which,

$$m = \tanh(\beta J m), \quad (1.31)$$

reproducing our earlier answer. This indicates that the mean-field approximation we made when we attempted to solve the short-range version of the model is exact in the case where interactions are long-ranged. (The missing factor of z arises because $z = N$ in a limit where every spin is connected to every other. This factor thus cancels with the $1/N$ required for the right thermodynamic limit.)

1.3.3 Mean-field Theory from the Saddle Point Approximation

Finally, it is also instructive to describe one further method to solve this problem. Consider

$$\mathcal{H} = -\frac{J}{2N}\left(\sum_i \sigma_i\right)^2. \quad (1.32)$$

Now
$$\mathcal{Z} = Tr_\sigma \exp\left[\frac{\beta J}{2N}\left(\sum_i \sigma_i\right)^2\right]. \qquad (1.33)$$

Recognizing this as a quadratic form, we can write this as,
$$\mathcal{Z} = \left(\frac{\beta N J}{2\pi}\right)^{1/2} \int_{-\infty}^{+\infty} dm\, Tr_\sigma \exp\left[-\frac{N\beta J}{2}m^2 + \beta J m \left(\sum_i^N \sigma_i\right)\right]. \qquad (1.34)$$

Taking the trace inside, we have
$$\mathcal{Z} = \left(\frac{\beta N J}{2\pi}\right)^{1/2} \int_{-\infty}^{+\infty} dm\, \exp\left[-\frac{N\beta J}{2}m^2\right] Tr_\sigma \exp\left[\beta J m \sum_i^N \sigma_i\right]. \qquad (1.35)$$

We can now perform the trace
$$\mathcal{Z} = \left(\frac{\beta N J}{2\pi}\right)^{1/2} \int_{-\infty}^{+\infty} dm\, \exp\left[-\frac{N\beta J}{2}m^2 + N\ln 2\cosh(\beta(Jm))\right]. \qquad (1.36)$$

which can be finally written as
$$\mathcal{Z} = \left(\frac{\beta N J}{2\pi}\right)^{1/2} \int_{-\infty}^{+\infty} dm\, \exp\left[-\beta J N f(m)\right], \qquad (1.37)$$

where
$$f(m) = \frac{1}{2}m^2 + \frac{1}{\beta J}\log[2\cosh\beta(Jm)]. \qquad (1.38)$$

Notice the way the system size N appears in the argument of the exponential. As $N \to \infty$, a saddle point evaluation of the integral becomes possible. The integral is dominated by its value at the minimum, located via
$$\frac{\partial f}{\partial m} = 0, \qquad (1.39)$$

which then gives us
$$m = \tanh\beta(Jm). \qquad (1.40)$$

This is identical to the simple expression evaluated earlier.

1.4 Landau-Ginzburg Theory of Phase Transitions

These ideas can be used to motivate a fundamental and intuitive approach to phase transitions, called Landau-Ginzburg theory after its founders. We start with the free energy derived earlier

$$\frac{F}{N} = -\frac{1}{2}Jm^2 + \frac{T}{2}[(1+m)\ln(1+m) + (1-m)\ln(1-m)] - T\ln 2, \qquad (1.41)$$

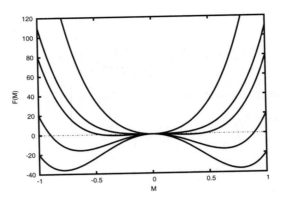

Figure 1.2: Landau Free energy as a function of the order parameter for a range of temperatures above and below the transition point. Above the transition temperature, this free energy has a minimum only at $M = 0$. Below the transition, there are minima at $\pm M_0$; the $M = 0$ state is a local maximum of the free energy.

which can now be expanded, using

$$(1+m)\ln(1+m) + (1-m)\ln(1-m) = m^2/2 + m^4/12, \tag{1.42}$$

to yield

$$\frac{F}{N} = -\frac{1}{2}(\beta J - 1)m^2 + \frac{T}{2}\left[m^4/12 + O(m^6)\ldots\right]. \tag{1.43}$$

Note that the free energy is minimized with $m = 0$ for temperatures greater than the critical temperature. For temperatures less than this critical temperature, the free energy is minimized with a non-zero value of m. A plot of this free energy for different temperature values both above and below the critical temperature, would look as illustrated in Fig. 1.2.

We can use this example to motivate a more general way of thinking about phase transitions. The order parameter is a new thermodynamic variable which enters the thermodynamic description of the model, below T_c. For magnetic systems which order ferromagnetically, the order parameter is the magnetization. For antiferromagnets, the order parameter is the staggered magnetization. Landau proposed that the dependence of the free energy on the order parameter could be modelled in a very general manner, with a form dictated by reasoning from symmetry arguments.

Consider a generic phase transition in which a quantity m is the order parameter and contemplate the general ingredients of an expansion of the free energy f in terms of m. Landau argued that an appropriate free energy density for the Ising problem could be obtained in a systematic expansion in the order

1.4. Landau-Ginzburg Theory of Phase Transitions

parameter. This form, as generalized by Landau and Ginzburg, is:

$$f = f_0 + am^2 + dm^4 + c(\nabla m)^2 \ldots \quad (1.44)$$

The assumptions which enter this are the following: There is always a non-singular part to the free energy density which we will lump into f_0. The $m \leftrightarrow -m$ symmetry of the model, in the absence of an externally applied field, ensures that only even powers of m should enter this expansion. We neglect all powers higher than 4, since we are interested in the vicinity of the critical point, where m is small. We assume that at large M the free energy should be monotonically increasing, so $d > 0$. We will see below that at $a > 0$ there is one phase, whereas for $a < 0$, there are two phases. Thus, a should change sign at $T = T_c$. At the simplest level, we can expand it as

$$a = \alpha(T - T_c). \quad (1.45)$$

If $c > 0$, the free energy can only increase if m is non-uniform. Thus, the full free energy is

$$F = F_0 + V\left[\alpha(T - T_c)m^2 + dm^4\right] \quad (1.46)$$

The minimum of the free energy gives,

$$2\alpha(T - T_c)m + 4dm^3 = 0, \quad (1.47)$$

with solutions

$$m = 0 \text{ or } m^2 = -\frac{\alpha(T - T_c)}{2d}. \quad (1.48)$$

This free energy can be used to calculate all the thermodynamic properties of the model.

If we define a **critical exponent** β which captures the increase of the magnetization m below T_c as

$$m \sim (T - T_c)^\beta, \qquad T < T_c, \quad (1.49)$$

then $\beta = 1/2$ within the Landau-Ginzburg theory. Our interest will be in the calculation of how similar thermodynamic functions become singular at the transition. Another example is the calculation of the magnetic susceptibility χ, which defines a related exponent γ i.e.

$$\chi \sim |T - T_c|^{-\gamma}, \quad (1.50)$$

where we anticipate that the value of γ is the same as the transition is approached from below or from above.

To obtain this exponent, we add a small external field H. This enters the Landau-Ginzburg expansion as

$$F = F_0 + V\left[\alpha(T - T_c)m^2 + dm^4 - Hm\right] \quad (1.51)$$

We want to calculate

$$\chi = \frac{\partial m}{\partial H}. \tag{1.52}$$

To obtain this, we use

$$2\alpha(T - T_c)m + 4dm^3 = H, \tag{1.53}$$

and distinguish two cases:

1. If $T > T_c$, then $m(H = 0) = 0$. For small H we neglect m^3, then

$$m = \frac{H}{2\alpha(T - T_c)} \tag{1.54}$$

and the susceptibility

$$\chi = \frac{1}{2\alpha(T - T_c)}. \tag{1.55}$$

This diverges at $T \to T_c$.

2. If $T < T_c$, $m(H = 0) = \pm m_0$ with $m_0 = \pm\sqrt{-\alpha(T - T_c)/2d}$. Let

$$m = m_0 + \delta m \tag{1.56}$$

$$\delta m = -\frac{H}{4\alpha(T - T_c)} \tag{1.57}$$

yielding

$$\chi = \mid \frac{1}{4\alpha(T - T_c)} \mid \tag{1.58}$$

Note that the susceptibility exponent is $\gamma = 1$, within Landau-Ginzburg theory.

The Ginzburg-Landau free energy can be used to derive predictions for the singular behaviour of a variety of observables, including the specific heat, the magnetization and others. The predictions here are identical to those obtained from the sorts of mean-field theories discussed here. Some of these observables diverge at the transition, with well-defined critical exponents, while quantities such as the specific heat are discontinous. This non-analytic behaviour at the transition, with associated algebraic dependence on the distance from the critical point, defines a number of **critical exponents**.

Our various derivations of mean-field theory and our brief description of Ginzburg-Landau theory illustrate several important aspects of the statistical mechanics of interacting systems:

- Mean field theory is exact as the number of neighbours goes to infinity. In general mean-field theory is a good qualitative guide to the physics.

1.4. Landau-Ginzburg Theory of Phase Transitions

- An order parameter is typically associated with a thermodynamic phase transition. The order parameter is usually defined to be zero on the high temperature side of the transition and non-zero below it.

- A common property of most (but not all) phase transitions is that a symmetry is broken in the low temperature phase: although the Hamiltonian may have a certain symmetry, the stable state at low-temperatures can break that symmetry. As an example, the ferromagnetic Ising model has two ground states and the system chooses one or the other just below the transition, in the thermodynamic limit.

- The barriers between possible symmetry broken states must become infinitely large, as the system size goes to infinity, *if the symmetry broken state is to be stable*. If this barrier were finite in the thermodynamic limit, the thermodynamic state would simply be an equal combination of the two symmetry broken states, since thermal fluctuations would always allow the surmounting of a finite barrier in a finite time.

- Several thermodynamic quantities behave in a singular manner at the phase transition point. Mean-field theory provides precise predictions, involving simple rational numbers, for this singular behaviour. These exponents, however, do not seem to be exact, in general, although they are believed to be exact in sufficiently high dimensions.

These general results will be useful in developing intuition for the statistical mechanics of disordered systems.

1.4.1 Critical Point Exponents

Scaling descriptions are common in statistical physics. Such descriptions are appropriate whenever the system behaves as though it lacks a "typical" or dominant scale for fluctuations. As an example, consider the Ising model. Repeated snapshots of configurations of the Ising model, say above its transition temperature, will show that fluctuations largely involve regions of a size *upto* a **correlation length** ξ. This means that such fluctuations will not typically involve regions of connected spins at a scale much larger than ξ.

Consider snapshots of configurations of an Ising model, say well below T_c. In this case, the system will largely align into either the all-up or all-down state. Thermal fluctuations allow some spins to flip opposite to the direction of mean alignment. Typically, the sizes of clusters of such spins - with two, three, four or more spins flipped opposite to the global alignment direction - is not very large, provided one is far away from the critical point. There is a well-defined scale (the correlation length, upto numerical factors) for such clusters, such that clusters well below such a scale are present in reasonable numbers, whereas clusters with sizes much larger than that scale are almost never seen. It will turn out that at scales smaller than the correlation length, the system behaves as if it has no intrinsic length scale, with the probability $P(s)$ of having a cluster of size s decaying as a power law with s. There are thus fluctuations of all sizes below this scale. As ξ increases upon approaching the transition, these regions of correlated spins become larger and larger, till the size of coherently fluctuating domains becomes comparable to the system size. There are thus fluctuations at all scales, ranging from those involving the flipping of one or a small number of spins from the aligned state to ones involving the flipping of a macroscopic number.

Behavior close to the critical point, in particular the non-analytic variation of thermodynamic functions, is of special interest to the statistical mechanic. Early work on understanding such behaviour led to the development of the concepts of **scale-invariance** and of **universality**. Many classes of apparently very disparate systems appear to have the same critical exponents, defining what are called **universality classes** of physical behavior. The reason for this is precisely the fact that correlation lengths diverge at continuous phase transitions. Universality results because microscopically different systems have the same large-scale description. As clarified by Wilson, Kadanoff and others in the development of the renormalization group, the essence of the critical point problem lies in the coupling of fluctuations over many scales, ranging from the microscopic to the macroscopic.

Mean-field theory provides one way in which the behaviour in the vicinity of the critical point can be calculated. Unfortunately, the predictions of mean-field theory are quantitatively wrong, although such theories are powerful ways of understanding statistical mechanical systems. Mean-field theory fails, at least for dimensions $d < 4$ in the short-ranged Ising case, because it ignores fluctuations, assuming that a single configuration dominates the partition sum. Mean-field behaviour is restored in models in high enough dimensions or if the interactions are sufficiently long-ranged.

Begin by defining a dimensionless scaled temperature which measures the distance from the critical point,

$$t = (T - T_c)/T_c. \tag{1.59}$$

In terms of this, if a quantity behaves as

$$F(t) \sim |t|^\lambda \tag{1.60}$$

close to the critical point, then the critical exponent λ can be obtained from

$$\lambda = lim_{t \to 0} \frac{\ln |F(t)|}{\ln t}. \tag{1.61}$$

Two-point correlation functions are defined through

$$G(r) \equiv \langle m(0)m(\mathbf{r}) \rangle. \tag{1.62}$$

Such a correlation function will generically decay as

$$G(r) \sim e^{-r/\xi}/r^{d-2}. \tag{1.63}$$

1.4. Landau-Ginzburg Theory of Phase Transitions

As one approaches the critical point, ξ the correlation length diverges as

$$\xi \sim |t|^{-\nu}. \tag{1.64}$$

Precisely at the critical point, the correlation function decays *asymptotically* as

$$G(r) \sim 1/r^{d-2+\eta}. \tag{1.65}$$

A full list of the conventionally defined critical exponents is the following:

$$\begin{aligned}
\text{Specific heat} \quad & C \sim |t|^{-\alpha} \\
\text{Magnetization} \quad & M \sim |t|^{-\beta} \\
\text{Isothermal susceptibility} \quad & \chi \sim |t|^{-\gamma} \\
\text{Critical isotherm} \quad & M \sim |H|^{-\delta} \\
\text{Correlation length} \quad & \xi \sim |t|^{-\nu} \\
\text{Correlation Function at } T_c \quad & G(r) \sim 1/|r|^{d-2+\eta}
\end{aligned} \tag{1.66}$$

We do not define different exponents for $T \to T_c^-$ and $T \to T_c^+$. The current understanding is that, where both exponents exist and are well defined, they are equal.

From thermodynamics, we can obtain various inequalities connecting the critical exponents.

$$\begin{aligned}
\alpha + 2\beta + \gamma &\geq 2 \\
\alpha + \beta(1+\delta) &\geq 2 \\
\gamma &\leq (2-\eta)\nu \\
d\nu &\geq 2-\alpha
\end{aligned} \tag{1.67}$$

These relations are believed to hold as equalities, a result which follows from the structure of the renormalization group approach. The last of these relations, due to Josephson, explicitly involves the dimension d and is called a **hyperscaling relation**. Such a relation is only valid at and below a special value for the dimension called the **upper critical dimension**. This critical dimension is 4 for the Ising case. At and above the upper critical dimension, the critical exponents are those predicted by mean-field theory.

In the table below, we tabulate the exponents as defined above. These exponents are (i) calculated within mean-field (Landau-Ginzburg) theory, (ii) obtained from an exact calculation for the two-dimensional Ising model due to Onsager, (iii) extracted from series expansions for the three dimensional Ising model, and finally, (iv), obtained from experiments on fluids near the liquid gas critical point.

Exponent	Value (Mean Field)	2d Ising Model	3d Ising Model	Experiments (Fluids)
α	0	0 (log)	0.109	0.15
β	0.5	1/8	0.327	0.33
γ	1.0	7/4	1.237	1.3
δ	3.0	15	4.78	4.5
ν	0.5	1	0.63	0.6
η	0.0	1/4	0.03	0.07

Note the similarity of exponents for the 3-d Ising model with those obtained from a variety of experiments on fluid systems near the critical point. This is no accident. Universality arguments indicate that these nominally very dissimilar systems belong to the same universality class and thus exhibit continuous transitions characterized by the same set of critical exponents.

1.5 The Percolation Problem and Geometric Phase Transitions

The **percolation** problem is a classic example of a geometrical phase transition. It is is exceedingly simple to pose. Consider a regular lattice, say square or triangular. Dilute this lattice by removing either sites (**site percolation**) or bonds (**bond percolation**). This is done with probability $(1-p)$ independently at each site. Thus, for large enough systems, Np sites or bonds are obtained on average. Obviously, once one has specified the lattice type, the only parameter of relevance is p.

Consider the behaviour of the bond percolation problem at both large and small p. For large p, very few bonds are removed. It is intuitively reasonable that, given any two sites on the lattice, a connected path exists from one to the other. In the opposite limit, where p is small, very few bonds survive and it is overwhelmingly likely that no such path should exist. It is reasonably to expect that these two asymptotic behaviours should be separated by a well-defined transition point. The transition can be described as a change in connectivity, where connecting two points whose asymptotic separation can be taken to be very large is either possible or impossible, on either side of the transition.

These arguments make it intuitively reasonable that a transition must exist. More quantitative arguments are required, however, to sharpen and quantify the argument. We do this by defining several quantities characterizing the connectivity in the different phases.

Let $n_s(p)$ be the number of clusters per lattice site of size s. The probability that a given site is occupied and belongs to a cluster of size s is thus $sn_s(p)$. Let $P(p)$ be the fraction of occupied sites belonging to the infinite cluster. Clearly, $P(p) = 1$ for $p = 1$ and $P(p) = 0$ for $p \ll p_c$. Thus $P(p)$ is analogous to an order parameter for a system which undergoes a thermal phase transition.

1.5. The Percolation Problem and Geometric Phase Transitions

We also have the relation

$$\sum_s sn_s(p) + pP(p) = p, \tag{1.68}$$

where the summation extends over all finite clusters.

One can also define the mean size of clusters, denoted by $S(p)$. This is related to $n(p)$ through the following:

$$S(p) = \frac{\sum_s s^2 n_s(p)}{\sum_s sn_s(p)}, \tag{1.69}$$

where the sum is again over finite clusters.

The pair connectedness $C(p, r)$ is the probability that occupied clusters a distance r apart belong to the same cluster. We can also define $G(p)$, the total number of finite clusters per lattice site, via

$$G(p) = \sum_s n_s(p). \tag{1.70}$$

This quantity is analogous to the free energy per site for magnetic phase transitions.

We now describe a scaling theory for the percolation problem. The percolation probability, or the probability that a given site belongs to the infinite cluster, behaves as

$$P(p) \sim (p - p_c)^\beta. \tag{1.71}$$

Similarly, we have

$$S(p) \sim (p_c - p)^{-\gamma} \tag{1.72}$$

$$G(p) \sim |p - p_c|^{2-\alpha}. \tag{1.73}$$

In percolation, the linear size of the finite clusters, below and above p_c, is characterized by the connectivity length ξ. This is similar to the correlation length in thermodynamic phase transition. Here ξ is defined as the root mean square distance between two sites on the same finite cluster, averaged over all finite clusters. It can be calculated by measuring the radius of gyration of a cluster

$$R_s^2 = \frac{1}{s^2} \sum_{i=1}^{s} |\mathbf{r_i} - \mathbf{r_0}|^2$$

where $\mathbf{r_0} = \sum_{i=1}^{s} \mathbf{r_i}/s$ is the position of the center of mass of the cluster and $\mathbf{r}_i$ is the position of the ith site of the cluster. The connectivity length is the average radius of gyration over all finite clusters and given by

$$\xi^2 = \frac{2\sum_s R_s^2 s^2 n_s(p)}{\sum_s s^2 n_s(p)}. \tag{1.74}$$

At p_c, clusters of all possible sizes, starting from a single site up to clusters spanning the system, appear. There are, correspondingly, very large fluctuations in cluster sizes. Due to the appearance of large finite clusters at the critical point, the connectivity length diverges as $p \to p_c$ with an exponent ν given by

$$\xi \sim |p - p_c|^{-\nu}.$$

The connectivity length exponent ν is also found to be related to the moment exponents of $n_s(p)$ via a hyper scaling relation. The infinite cluster at $p = p_c$ contains holes of all possible sizes; the percolation cluster is self similar and fractal. A hyperscaling relation $\gamma + 2\beta = d\nu$ is satisfied between the scaling exponents. The pair connectedness function behaves as:

$$C(p, r) \sim \frac{\exp -r/\xi(p)}{r^{d-2+\eta}}. \tag{1.75}$$

The scaling theory assumes that for p near p_c, there is a typical cluster size s_ξ which leads to the dominant contribution to these divergences. Since this diverges as $p \to p_c$, we can assume that

$$s_\xi \sim |p - p_c|^{-1/\sigma} \tag{1.76}$$

We will also assume that

$$n_s(p) = n_s(p_c) f(\frac{s}{s_\xi}) \tag{1.77}$$

We require that the function $f(x) \to 0$ as $x \to \infty$ and $f(x) \to 1$ as $x \to 0$. We know that, for large s, the scaling form $n_s(p_c) \sim s^{-\tau}$ is obtained. Thus we get

$$n_s(p) = s^{-\tau} \phi(s|p - p_c|^{1/\sigma}) \tag{1.78}$$

The percolation exponents can therefore be obtained in terms of two independent exponents σ and τ.

The percolation problem can be solved exactly in one dimension. It can be easily seen that, for long range connectivity, all the sites must be present, which implies that $p_c = 1$. For clusters containing s-sites: $n_s = p^s(1-p)^2$. The $(1-p)^2$ term comes from the probability of two empty sites that bind the cluster from two ends. Once this distribution is known it is easy to obtain the behavior of various observables near $p_c = 1$. The values of the critical exponents can be obtained by taking suitable moments of the cluster size distribution as stated before.

In two dimensions, the exponents of the percolation problem are given in terms of rational numbers. The exponents are listed below in table 1.1.

The relationship between thermal phase transitions and percolation was first discovered by Fortuin and Kastelyn who showed that the bond percolation problem could be mapped to the q–state Potts model in the limit that $q \to 1$. This mapping can be used to obtain the critical exponents in the equations above.

The values of the critical exponents are found to be same on different lattices and for site or bond percolation. The exponents depend only on the space dimension, another illustration of universality.

1.6. Disordered Spin Systems

Dimension	d=2	d=3	mean-field
β	5/36	0.41	1
γ	43/18	1.80	1
ν	4/3	0.88	1/2
α	−2/3	-0.62	-1
σ	36/91	0.45	1/2
τ	187/91	2.18	5/2

Table 1.1: The values of the critical exponents of the percolation phase transition.

1.6 Disordered Spin Systems

There are many ways in which disorder can enter the specification of a simple model such as the ones we have been studying. For one, some of the spins could be absent from their regular sites. For another, the interaction between two spins could depend on the labels of the spins, with $J \to J_{ij}$. A third possibility is the presence of random fields acting on the spins independently.

Generalize the model Ising Hamiltonian to the form

$$\mathcal{H} = -\sum_{\langle ij \rangle} J_{ij} \sigma_i \sigma_j - \sum_i h_i \tau_i \qquad (1.79)$$

where the disorder is contained in the variables J_{ij} and the random fields h_i. The problem in which all the h_i's are set to zero and the J_{ij}'s are random is the **random exchange model**. The problem with $J_{ij} = J$ i.e. *uniform* exchange couplings but with h_i random is the **random field model**. In general, in experimental systems, both types of disorder could be present. One could also have situations where spins are missing from sites, the case of **site diluted magnets** or where the magnitude of the spin could differ from site to site. Further, instead of considering Ising spins, one could consider spin variables which take values on a circle (XY spins) or on the surface of a sphere (Heisenberg spins).

How does one perform the average over quenched variables such as J_{ij}? The correct way of thinking about this is the following: Imagine subdividing the macroscopic system into many sub-parts, each large enough to be treated macroscopically, as in Fig. 1.3. The interaction between each subsystem is a surface effect, one which is subleading with respect to the bulk in the limit of large subsystem size. Note that the free energy in each subsystem is a well-defined quantity but fluctuates from subsystem to subsystem since the microscopic configuration of disorder is different in each subsystem. We may thus obtain the free energy and thermodynamic functions within each subsystem, in principle, by doing the partition function sum keeping the quenched variables fixed. We must then decide how to average over disorder, in order not to have tp provide a detailed microscopic specification of the disorder variables. This is done using the following argument. Each subsystem provides an independent realization

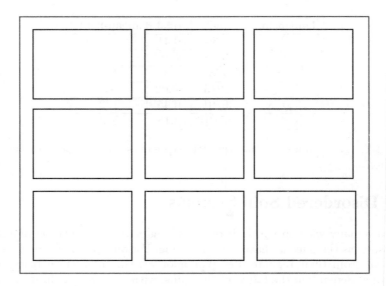

Figure 1.3: The decomposition of a thermodynamic system into subsystems

of disorder, derived from the probability distribution for the quenched variables. In the limit of an infinite number of subsystems whose interactions can be ignored to a first approximation (this is a surface and not a bulk term), the disorder average is the average over the subsystems. Averaging over disorder is equivalent to averaging over subsystems for variables such as the free energy whose values can be appropriately defined by summing up over subsystems.

A quantity is called **self-averaging** if its distribution over the subsystems as described above is not anomalously broad. This is equivalent to requiring that the mean-value and the most probable value coincide in the thermodynamic limit. Note that the partition function is *not* self-averaging and cannot be averaged in the same way as the free energy; it does not decompose naturally into a bulk and sub-leading surface term. Checking whether a given thermodynamic quantity is properly self-averaging or not is important, since the statistical averaging may otherwise yield mean values that are not the values measured in a typical experiment. The argument above, due initially to Brout, indicates that the free-energy and similar observables are self-averaging, for interactions which are short-ranged and for probability distributions over disorder which are "well-behaved".

A useful example of the distinction between typical and average is the following: Consider a random variable x that takes two values

$$X_1 = e^{\alpha\sqrt{N}} \text{ and } X_2 = e^{\beta N}, \beta > 1, \tag{1.80}$$

with probabilities

$$p_1 = 1 - e^{-N}, \text{ and } p_2 = e^{-N}. \tag{1.81}$$

1.6. Disordered Spin Systems

In the limit $N \to \infty$, the average value $\to e^{(\beta-1)N}$ while the typical or most probable value is $x = X_1$ with probability 1. On the other hand the average value of $\ln x \to \alpha\sqrt{N}$ in the same limit, showing that the most probable value is determined by the typical value of the variable while the moments are controlled by the rare events. Provided x has a probability distribution which does not have special or rare events which occur with low probability but contribute anomalously to mean values and higher moments, this problem of matching the average value with the most probable value does not exist.

Quenched disorder however is not the only ingredient required for non-trivial statistical mechanics. The other ingredient is **frustration**, and it is the combination of disorder and frustration which leads to the remarkable properties of spin-glass systems. To understand frustration, consider a model in which

$$\mathcal{H} = -\sum_{\langle ij \rangle} \tau_i \tau_j \sigma_i \sigma_j, \qquad (1.82)$$

where we have taken the special case in which $J_{ij} = \tau_i \tau_j$, and $\tau_i = \pm 1$. We assume that the τ's are genuine quenched variables. Nevertheless all of the statistical mechanical content of this model (*nota bene*, in the absence of a magnetic field) can be shown to be trivially related to the pure version of the Ising model. To do this, simply redefine

$$\sigma'_i = \tau_i \sigma_i. \qquad (1.83)$$

The partition function now reduces to

$$\mathcal{Z} = Tr_{\sigma'_i} e^{-\beta J \sum_{\langle ij \rangle} \sigma'_i \sigma'_j}. \qquad (1.84)$$

This is, however, just the pure Ising model in another guise, since disorder now does not appear explicitly in the specification of the partition function, while all spins are summed over. This model, the **Mattis model**, provides an example of the way in which, in some cases, disorder can be transformed away. What remains is an effective pure problem in a redefined set of variables.

There is an interesting peculiarity of the transformation of spin degrees of freedom discussed in the previous section. If we multiply the spin variables τ_i across any closed circuit, the answer is always 1. This is easily traced to the fact that each τ_i appears twice when multiplied along a closed loop. Relating this to our discussion of the Mattis model above, what this means in practice is that the ground state of the model is non-degenerate, upto a global spin flip, since one can always invert the transformation above to obtain the original spin variables associated with the minimum energy configuration. Since all the variables $\tau_i \sigma_i$ must be equal in the ground state, the spin values which minimize the energy can be trivially extracted.

Models where this does not happen are frustrated models. The Mattis model exhibits disorder but not frustration. The idea of frustration can be

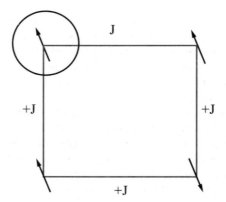

Figure 1.4: Illustration of Frustration: The circled spin receives conflicting information from its neighbours and can point either up or down with the same energy. There is no way of minimizing its energy.

illustrated by considering spins on a square block with nearest neighbor interaction as shown in Fig. 1.4. Take the spins on the square. While the energy of any pair of spins can be appropriately minimized by minimizing the bond energy, one spin indicated by the circle in the figure receives contradictory instructions from the spins it is bonded to, since it is bonded ferromagnetically to one and antiferromagnetically to another. Thus not all interactions can be satisfied automatically.

Frustration refers to the impossibility of satisfying all interactions at the same time to attain an absolute minimum of the free energy. The combination of both quenched disorder and frustration are essential ingredients of non-trivial behaviour in disordered systems. It is easy to come up with examples of systems which exhibit frustration in the absence of quenched disorder (such as the Mattis model) and systems which exhibit quenched disorder in the absence of frustration, such as non-interacting particles in a disordered background. Frustration and disorder are essential ingredients of all models of spin glasses, since the combination of these two yield the most important property of spin-glass materials, the larger number of low-lying (or degenerate) free energy minima and extremely complex patterns of ergodicity and symmetry breaking.

There are three basic scenarios for spin glass-like behaviour. In the first one, barriers separating low free energy minima remain finite at all nonzero temperatures. Thus the relaxation of the system may be anomalous but the system remains paramagnetic. In the second, it could happen that there is a certain spin state, or a finite number of such states related by symmetry, with a lower free energy than any other. In such a frozen state, an appropriate order parameter describing the freezing of the spins along random directions, the Edwards-Anderson order parameter to be defined later, would be non-zero. This case can still be studied along traditional lines. In the last scenario, we

1.6. Disordered Spin Systems

could have an extensive number of ground states with the system frozen in any one of these states. The infinite-ranged spin glass model to be discussed in the following sections corresponds to the last case.

1.6.1 The Harris Criterion

When does the presence of disorder change the universality class of a transition? This problem was considered some decades ago by Harris. Harris argued the following: Consider a system where the disorder couples to the temperature, say through a variable p. Thus, different regions of the sample have different critical temperatures $T_c(p)$. We may thus ask: Does allowing the local T_c to vary in this manner alter the critical exponents of the pure transition or leave them unchanged?

Begin with the pure system. Let the correlation length ξ be large. The volume of a coherence region is

$$V_\xi \sim \xi^d, \tag{1.85}$$

in d dimensions. Since the correlation length

$$\xi \sim t^{-\nu}, \tag{1.86}$$

we have

$$V_\xi \sim t^{-\nu d}. \tag{1.87}$$

relating the size of the coherence volume to the distance from the transition. The t variable can then be calculated from

$$t \sim V_\xi^{-1/\nu d}. \tag{1.88}$$

Now disorder gives rise to fluctuations of the local transition temperature. Given a volume V, this fluctuation, from central limit theorem arguments relying on the uncorrelated character of the disorder, should scale as the square root of the number of degrees of freedom in that volume. Thus, disorder-induced fluctuations away from the true T_c scale as $V_\xi^{1/2}$. Comparing terms, we get $d\nu < 2$ as a condition for disorder-induced fluctuations to dominate, driving the system away from the critical fixed point corresponding to the pure system.

We can now make use of the hyperscaling relation

$$d\nu = 2 - \alpha. \tag{1.89}$$

This yields,

$$\alpha > 0, \tag{1.90}$$

as the condition for disorder to be a relevant perturbation, a result first obtained by Harris. Note that, subject to certain mild conditions, for a sharp phase transition in a disordered system to occur, one must necessarily have $\alpha < 0$. That is, the specific heat must have the form of a cusp.

1.6.2 The Random Field Problem and the Imry-Ma argument

Consider a model with the Hamiltonian

$$\mathcal{H} = -J\sum_{\langle ij \rangle} \sigma_i \sigma_j - \sum_i h_i \sigma_i. \tag{1.91}$$

Note that the exchange interaction is uniform, but that disorder enters in the form of a random field coupling independently to spins at each site. The spins above are Ising spins. They could, in general, be $n-$component spins, with $n = 1$ the Ising case, $n = 2$ the **XY model** and $n = 3$ the **Heisenberg model**. The pure problem has a transition with a lower critical dimension of 1 (Ising) and of 2 ($n \geq 2$). The question of whether the transition survives the introduction of weak random-field disorder can be posed in terms of the stability of the ground state.

Imry and Ma formulated a simple yet elegant argument to investigate such ground state stability. We illustrate this for the Ising case. Consider the aligned ground state in the absence of a random field. Now introduce the random field. Over a given region of dimension L, the field is, on average zero, but fluctuates about this average. Thus the value of the $h_i \sigma_1$ term, fluctuates to order $L^{d/2}\Delta$, where Δ denotes the width of the field distribution. If this fluctuation is such that the system can lower its energy by flipping all spins in that region, it will do so, while paying a cost for creation of an interface at the boundary. If a single length L dominates, then this interface cost is $\propto L^{d-1}$ in $d-$dimensions (the Ising case).

Combining these two terms into a free energy cost for flipping the domain yields the two competing terms

$$\Delta F = -\Delta L^{d/2} + J L^{d-1}. \tag{1.92}$$

The minimum of the free energy can be obtained as

$$\frac{\partial F}{\partial L} = 0 \implies -\frac{d}{2}\Delta L^{d/2-1} + J(d-1)L^{d-2}. \tag{1.93}$$

This then yields

$$L^* \sim [J/\Delta]^{2/(2-d)} \tag{1.94}$$

One can see that a change in behaviour occurs when $d \geq 2$, For dimensions $d < 2$, there is a finite L^* at which the energy gain from flipping the domain outweighs the energy cost of creating a domain wall. It is always favourable to flip domains of dimension larger than L^*, given the energy gain from the random field. For dimensions $d > 2$, the cost of creating the domain wall dominates at large scales and only a few small domains are formed, insufficient in number to fully destroy the long-ranged order in the ground state of the pure model.

This argument can be generalized to the case of continuous spins, with the only modification required relating to the domain wall energy term. Since in

1.6. Disordered Spin Systems

continuous spin systems, the energy cost of the domain wall can be spread out over a region of size L, the ordered ground state is stable for $d > 4$ and unstable below that. A rigorous proof of the lower critical dimension $d_c = 2$ for the random field Ising model was provided by Spencer. A detailed discussion of the physics of random field models, extending and clarifying the simple argument above, is contained in the chapter of P. Shukla.

1.6.3 Griffiths Phase

A remarkable result for dilute magnetic systems was obtained by Griffiths, who considered simple Ising spin systems on a lattice "diluted" through the random removal of a certain fraction of sites. If a sufficiently large fraction of sites are removed, sites will not percolate and the system essentially breaks up into a number of independent domains. In such a state, there can be no long range order, since each domain can flip independently and the size of each domain is finite.

Griffiths noted that the situation of independent domains lacking long-range order could nevertheless exhibit unusual behaviour in the thermodynamic limit. He examined the zeros of the partition functions for finite sized systems of volume V. These zeros, the Yang-Lee zeros, lie on a circle of unit radius in the complex plane. The transition in the pure system in the thermodynamic limit is related to the fact that these zeros pinch the real axis in the thermodynamic limit, showing that the high temperature and low temperature phases in the thermodynamic limit are not analytically connected. This analysis provides a simple way of understanding the phase transition in the pure system.

Griffiths showed that these zeros, for the finite clusters obtained in the diluted system, approached the real axis arbitrarily closely for larger volumes. Griffiths proved that although due to the almost-sure finiteness of these clusters no first-order transition with ferromagnetic long-range order can occur, in the paramagnetic phase below the critical temperature of the non-dilute Ising model the free energy and magnetisation are nevertheless non-analytic as a function of the magnetic field h at the value $h = 0$. This result is remarkable because it demonstrates that it is possible to have a range of temperatures over which non-analyticity occurs as opposed to a single T_c, it shows that there may be nonanalytic behavior even if the correlation length is finite and, finally, it shows that singularities other than power laws may be a generic possibility in disordered systems.

Specifically, a dilute ferromagnet is in the **Griffiths phase** if its temperature T is between the critical temperature $T_c(p)$ for the onset of magnetic long-range order and the critical temperature T_c of the nondilute system. More generally, a random magnetic system is in its Griffiths phase if it is above its own ordering temperature but below the highest ordering temperature allowed by the distribution of disorder. In the Griffiths phase the free energy exhibits essential singularities. Physically, this singular behaviour is due to the occurrence, with low but finite density in the infinite volume, of arbitrarily large rare regions, defined in the following way: In an infinite system one can find arbitrarily large spatial regions that are devoid of impurities. For temperatures between $T_c(p)$ and T_c, these regions will show local magnetic order even though the bulk system is globally in the paramagnetic phase. These spatial regions are known as **rare regions** and fluctuations involving these regions are non-perturbative degrees of freedom excluded in conventional approaches to phase transitions based on perturbation theory. In classical systems with uncorrelated disorder, thermodynamic Griffiths effects are very weak as a consequence of the essential singularity. The magnetic susceptibility remains finite within the Griffiths region. In contrast, the long-time spin dynamics inside the Griffiths phase is dominated by the rare regions, leading to very slow relaxation. Griffiths singularities assume added importance in the context of quantum disordered systems.

1.7 Spin Glass Basics: Experimental Phenomenology

Studies of the spin glass state were initially motivated by the desire to understand the low-temperature behaviour of certain magnetic alloys. While early work concentrated on understanding the fate of isolated magnetic impurities in a non-magnetic (typically metallic) host, it was realized that understanding the interactions between impurities was a challenging problem. Such systems of dilute magnetic impurities coupled to each other behaved as though there was a low-temperature freezing of the moments associated with the magnetic ions. The field of spin glasses emerged from these early studies.

There are a variety of different experimental systems which exhibit spin glass behaviour: (I) stable spins (Mn, Fe Gd, Eu) diluted into non-magnetic metals (Cu,Au) such as $Cu_{1-x}Mn_x$, $Au_{1-x}Fe_x$, (II) nonmagnetic compounds with some magnetic ions, such as $Eu_xSr_{1-x}S$ or La_{1-x} Ga_x Al_2 or (III) amorphous intermetallics such as $GdAl_2$, YFe_2. Spin glass samples exhibit irreversible behaviour below T_f and reversible behaviour above it: cooling down in constant field (field-cooled or FC) leads to smooth reversible curves below T_f while zero-field cooling (ZFC) leads to irreversible behaviour which relaxes towards FC behaviour very slowly. Typical plots are shown in Fig. 1.5.

The principal experimental signatures of spin glasses are the following: the specific heat at zero applied field shows a broad maximum at about 1.4 T_f, coupled with a slow decrease at high T. The peak is smeared by the application of a field. The dc susceptibility exhibits a sharp cusp at T_f. This cusp is very strongly affected by magnetic fields, with even very small fields capable of rounding out the cusp, as in Fig. 1.6. The ac magnetic susceptibility peak shows a shift depending on frequency, in contrast to behaviour in conventional ferromagnets or antiferromagnets. There is evidence for a large distribution of relaxation times in ac susceptibility. Typically there are no clear signs of freezing in resistivity. Over 500 different experimental systems exhibiting spin glass behavior have now been identified, including transition-metals in

1.7. Spin Glass Basics: Experimental Phenomenology

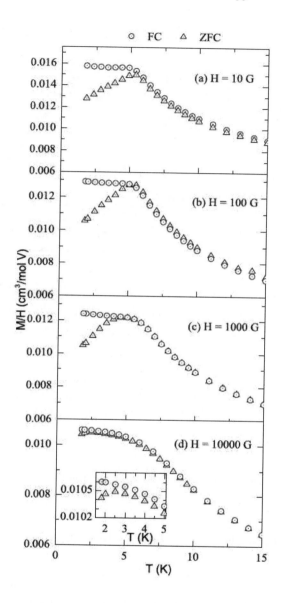

Figure 1.5: Splitting of field cooled (FC) and zero-field cooled (ZFC) magnetization with temperature of $(Li_{0.40}V_{0.60})_3B\,O_5$ crystals at low temperatures and fields of $H = 10, 100, 1000$ and 10000 Gauss. The inset is an expanded plot of the low temperature region. Figure reproduced with permission from X. Zong, A. Niazi, F. Borsa, X. Ma and D.C. Johnston, Phys. Rev. B **76** 054452 (2007). Copyright (2007) by the American Physical Society.

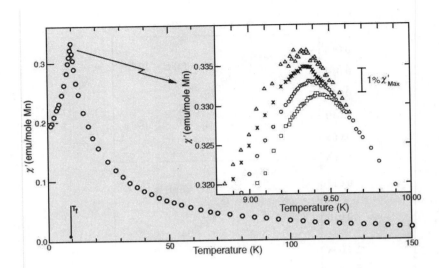

Figure 1. AC susceptibility of CuMn (1 at% Mn) showing the cusp at the freezing temperature. The inset shows the frequency dependence of the cusp from 2.6 Hz (triangles) to 1.33 kHz

Figure 1.6: Illustration of the cusp in the real part of the ac susceptibility, indicating the spin glass transition and its broadening with increasing frequency. Figure reproduced with permission from C.A.M Mulder, A.J. van Duyneveldt and J.A. Mydosh, *Phys. Rev. B* **23** *1384 (1981)*2007. Copyright (1981) by the American Physical Society

noble metal solutes, rare earth combinations, amorphous metallic spin glasses, semiconducting spin glasses and insulating spin glasses.

1.8 Spin Glass Physics: Towards a Model

To understand the spin glass problem, we must understand the fate of magnetic impurities in metals. Indirect exchange, mediated through conduction electrons in metallic systems, couples moments over large distances. It is the dominant exchange interaction in metals where there is little or no direct overlap between neighbouring magnetic electrons.

In a conventional metallic system with dilute magnetic impurities, the impurities are far apart. They can thus be modeled by what are called single-impurity Anderson or Kondo Hamiltonians, modeling the problem of an isolated magnetic impurity coupled to a sea of conduction electrons. The properties of such a system are by now fairly well understood at a quantitative level. When

1.8. Spin Glass Physics: Towards a Model

the density of magnetic impurities is high, however, we can no longer ignore spin-ordering tendencies arising from the interaction between impurities. The two competing effects that lead to two different tendencies of the system to interact with conduction electrons, known as the Kondo effect and the Ruderman-Kittel-Kasuya-Yoshida (RKKY) interaction, have a crucial influence on impurity magnetism. The conduction electron-mediated, RKKY indirect-exchange interaction favors magnetic ordering whereas the Kondo effect tends to quench individual impurity spins.

Hybridization between the conduction electrons of the host metal and the d- (or f-) electrons of the magnetic impurity produces an effective on-site exchange coupling at the impurity site. The sign of the interaction is typically antiferromagnetic, as the conduction electrons attempt to screen the spin of the impurity in their midst. Rather than forming a negative spin-1/2 at the impurity site, however, the electrons spin-polarize in concentric rings around the impurity. This leads to an interaction of the form

$$J(\vec{R}_i - \vec{R}_j) \sim \cos(2k_F R)/R^3. \tag{1.95}$$

with k_f the Fermi wave-vector. This is the RKKY interaction.

The source of the rings of alternating polarization is that a true δ-function in space would require, in Fourier space, all k-vectors from 0 to infinity to be equally weighted, Since the host is a metal, available k−vectors range only from 0 to k_F. The system thus cannot screen the impurity spin in a localized manner. Instead, it first creates an opposite alignment in the vicinity of the impurity which overscreens it, followed by a parallel alignment further away which overcompensates in the opposite direction. This proceeds with decreasing amplitude out to infinity. Thus, depending on the distance between two impurity spins, their coupling could be either ferromagnetic or anti-ferromagnetic, resulting in frustration.

The most realistic model for spin glasses would consider vector spins occupying random sites. The Hamiltonian would then read

$$\mathcal{H} = \sum_{\langle ij \rangle} J(\vec{R}_i - \vec{R}_j) \vec{S}_i \cdot \vec{S}_j. \tag{1.96}$$

The origins of this magnetic interactions can be several, including direct exchange, RKKY, superexchange and dipolar.

A simpler model for freezing retains some of these ingredients: Only RKKY interactions are considered, there is no Kondo effect or long-range order and one considers only the interactions of single spins. The freezing temperature can be estimated in the following way. The RKKY interaction energy is of the form: $J_{RKKY}(r) \sim J_0/r^3$. The average impurity distance scales as $(1/x)^{1/3}$ with x the impurity concentration. Given a thermal energy whose scale is set by $k_B T$, the freezing temperature is then obtained by equating $J_{RKKY} = k_B T_f$, which implies that $T_f = x J_0 / K_B T$.

In the Edwards-Anderson model, the combination of site disorder and the RKKY interaction gives a random set of bonds satisfying a Gaussian distribution. Edwards and Anderson considered a non-dilute system of spins with such random bonds, describing the spin freezing in terms of each spin locking on to a random direction. If we think of the orientation of the spin vector at time t as $\sigma_i(t)$, Edwards and Anderson proposed the following order parameter, illustrated here for the Ising case:

$$q = \lim_{t\to\infty} q(t) = \lim_{t\to\infty} \langle\sigma(0)\rangle\langle\sigma(t)\rangle. \qquad (1.97)$$

This definition involves the projection of the thermally averaged value of a spin at a site i at a particular time with its value at an infinite time later. (If the spin is frozen in orientation, this projection yields a non-zero result. Note that this particular order parameter would be non-zero even for a ordered ferromagnet. However, the definition of the canonical spin-glass state has an additional ingredient - the average magnetization is required to also vanish, which it does not do in the ordered ferromagnet.)

This was a dynamic definition of the spin-glass order parameter, since it explicitly involved time. An ensemble-based description takes the form

$$q = \frac{1}{N}\left[\langle\sigma_i\rangle\langle\sigma_i\rangle\right] \qquad (1.98)$$

where the brackets $[\cdot]$ denote a disorder average and the brackets $\langle\cdot\rangle$ denote a thermal average.

The Edwards-Anderson method involved the first use of what is now called the replica method or replica "trick"', which we will discuss in some detail later. The model gave good predictions for susceptibility but disagreed with the behaviour of the specific heat seen in experiments. Later work by Thouless, Anderson and Palmer described a "cavity mean-field theory" which showed the existence of a macroscopic number of metastable states within this model.

1.9 The Sherrington Kirkpatrick Model

The model we discuss here is called the Sherrington-Kirkpatrick model. It is a model with infinite-ranged interactions, in which the coupling of two spins i and j depends on the labels i and j but not on the locations of the spins in physical space. The pure version of the SK model was solved exactly, by saddle point techniques, in an earlier section.

We begin with the Hamiltonian. The spins σ_i are Ising spins, taking values ± 1. The sum is over all pairs of spins and the model is infinite-ranged as a consequence.

$$\mathcal{H} = -\sum_{i\langle j} J_{ij}\sigma_i\sigma_j. \qquad (1.99)$$

We do not introduce factors of $1/N$ as we did previously. Such factors will be included in the definition of the distribution for the J_{ij}'s.

1.9. The Sherrington Kirkpatrick Model

The J_{ij}'s are quenched random interactions generated *via* the probability distribution

$$P[J_{ij}] = \prod_{i<j} \sqrt{\frac{N}{2\pi}} \exp\left(-\frac{J_{ij}^2 N}{2}\right). \qquad (1.100)$$

This distribution, being Gaussian, is completely specified by its mean and standard deviation

$$[J_{ij}] = 0, \, [J_{ij}^2] = \frac{1}{N}. \qquad (1.101)$$

Averages with respect to this probability distribution are defined through

$$[\cdot] = \int \mathcal{D}J P[J](\cdot) = \prod_{i<j} \sqrt{\frac{N}{2\pi}} \int_{-\infty}^{+\infty} \exp\left(-\frac{J_{ij}^2 N}{2}\right)(\cdot). \qquad (1.102)$$

We would like to calculate the disorder-averaged free energy which encodes all thermodynamic properties of this model *i.e.*

$$F = [F_d], \qquad (1.103)$$

where the brackets $[\cdot]$ denote an average over the probability distribution for the quenched variables, the exchange terms J_{ij}'s. Of course the difficulty lies precisely in the fact that a *logarithm* must be averaged. This was circumvented, in a technique invented for this purpose in the original Edwards-Anderson paper, where the following expansion was used

$$Z^n = \exp n \ln Z = 1 + n \ln Z + \ldots \qquad (1.104)$$

and thus

$$\ln Z = \lim_{n \to 0} \frac{Z^n - 1}{n}. \qquad (1.105)$$

Note that now what is required is the average of a power of the partition function. This is certainly easier to do than averaging the logarithm. We can now write

$$Z^n = \prod_{\alpha=1}^{n} Tr_{\sigma_\alpha} \exp \sum_{\alpha=1}^{n} \sum_{i<j}^{N} J_{ij} \sigma_i \sigma_j, \qquad (1.106)$$

from which the free energy follows as

$$F_n = -\frac{1}{\beta} \ln Z_n, \qquad (1.107)$$

and we have performed the average

$$Z_n \equiv [Z_J^n]. \qquad (1.108)$$

Now, taking the limit yields

$$\lim_{n \to 0} F_n = -\frac{1}{\beta}[\ln Z_J] = F. \tag{1.109}$$

We now return to the explicit calculation of the nth power of the partition function

$$Z_n = (Tr_{\sigma_\alpha^i}) \int \mathcal{D}J_{ij} \exp\left\{\beta \sum_{\alpha=1}^{n} \sum_{i<j}^{N} J_{ij}\sigma_i^\alpha \sigma_j^\alpha - \frac{1}{2}\sum_{i<j} J_{ij}^2 N\right\}. \tag{1.110}$$

Note that we can write this as

$$Z_n = (Tr_{\sigma_\alpha^i}) \int \mathcal{D}J_{ij} \exp\left\{\beta \sum_{i<j}^{N} J_{ij} \sum_{\alpha=1}^{n} \sigma_i^\alpha \sigma_j^\alpha - \frac{1}{2}\sum_{i<j} J_{ij}^2 N\right\}, \tag{1.111}$$

interchanging the order of the summations. Then, averaging over disorder yields

$$Z_n = (Tr_{\sigma_\alpha^i}) \exp\left[\frac{\beta^2}{2N}\sum_{i<j}^{N}\left(\sum_{a=1}^{n}\sigma_i^a\sigma_j^a\right)^2\right]. \tag{1.112}$$

We will attempt to simplify the term in

$$Z_n = \sum_{i<j}^{N}\left(\sum_{a=1}^{n}\sigma_i^a\sigma_j^a\right)^2. \tag{1.113}$$

Fixing $\{ij\}$, we write each term as

$$\sigma_i^1\sigma_j^1\sigma_i^1\sigma_j^1 + \sigma_i^1\sigma_j^1\sigma_i^2\sigma_j^2 + \sigma_i^1\sigma_j^1\sigma_i^3\sigma_j^3 + \sigma_i^2\sigma_j^2\sigma_i^2\sigma_j^2 + \ldots \tag{1.114}$$

There are n terms (for n replica's in this expansion) which contributes unity from terms which involve the same replica index; there are $N^2/2$ terms arising from the sum over ij arising out of all possible pairs. So we can pull this out to get

$$\exp\left[\frac{\beta^2}{2N} \times n \times \frac{N^2}{2}\right]. \tag{1.115}$$

Now we consider the expansion of the terms left out in doing this

$$\sigma_i^1\sigma_j^1\sigma_i^2\sigma_j^2 + \sigma_i^1\sigma_j^1\sigma_i^3\sigma_j^3 + \sigma_i^2\sigma_j^2\sigma_i^3\sigma_j^3 + \ldots \tag{1.116}$$

We reorder these as

$$(\sigma_i^1\sigma_i^2)(\sigma_j^1\sigma_j^2) + (\sigma_i^1\sigma_i^3)(\sigma_j^1\sigma_j^3) + (\sigma_i^2\sigma_i^2)(\sigma_j^3\sigma_j^3) + \tag{1.117}$$

1.9. The Sherrington Kirkpatrick Model

These are terms which turn up in the expansion of

$$\left[\sum_i (\sigma_i^a \sigma_i^b)\right]^2. \tag{1.118}$$

Now both $a > b$ and $b > a$ appear in the sum above; by symmetry of the terms we can consider just one, say $a < b$ and multiply by a factor of 2. Also, while the sum outside is over $i < j$, we may divide by 2 and retain the term. So factors of 2 cancel from top and bottom. Now we have

$$\mathcal{Z}_n = \sum_{\sigma_i^a} \exp\left[\frac{1}{4}\beta^2 N n + \frac{\beta^2 N}{2} \sum_{a<b}^n \left(\frac{1}{N}\sum_i^N \sigma_i^a \sigma_i^b\right)^2\right] \tag{1.119}$$

We restrict the sum to $a < b$ because we have already accounted for the $a = b$ terms by pulling them out. Following the same procedures we used earlier in the mean-field theory for the pure problem

$$Z_n = \prod_{a<b}^n \left(\int dQ_{ab}\right) \sum_{\sigma_i^a} \exp\left(\frac{1}{4}\beta^2 N n - \frac{\beta^2 N}{2}\sum_{a<b}^n Q_{ab}^2 \right. \tag{1.120}$$

$$\left. + \beta^2 \sum_{a<b}^n \sum_1^N Q_{ab}\sigma_i^a \sigma_i^b\right).$$

Here, the quantities Q_{ab} are related to the correlation functions

$$Q_{ab} = \frac{1}{N}\sum_{i=1}^N \langle \sigma_i^a \sigma_i^b \rangle. \tag{1.121}$$

Then we have, upon taking the sum over spins inside, and using the fact that now the sites are decoupled from each other,

$$Z_n = \prod_{a<b}^n \left(\int dQ_{ab}\right) \exp\left(\frac{1}{4}\beta^2 N n\right. \tag{1.122}$$

$$\left. - \frac{\beta^2 N}{2}\sum_{a<b} Q_{ab}^2\right) \prod_i^N \sum_{\sigma_i^a} \exp\left(\beta^2 Q_{ab}\sigma_i^a \sigma_i^b\right). \tag{1.123}$$

The second product involves N independent terms. The value of each term is

$$\left[\sum_{\sigma_a} \exp \beta^2 \sum_{a<b}^n Q_{ab}\sigma_a \sigma_b\right]. \tag{1.124}$$

This can be exponentiated to yield

$$Z_n = \prod_{a<b}^{n} \left(\int dQ_{ab} \right) \sum_{\sigma_i^a} \exp \left(\frac{1}{4} \beta^2 N n - \frac{\beta^2 N}{2} \sum_{a<b} Q_{ab}^2 \right. \quad (1.125)$$

$$\left. + N \log \left[\sum_{\sigma_a} \exp \beta^2 \sum_{a<b}^{n} (Q_{ab} \sigma_a \sigma_b) \right] \right).$$

Note that this can now be represented as

$$Z_n = \int d\hat{Q} \exp\left(-\beta n N f\left[\hat{Q}\right]\right). \quad (1.126)$$

where

$$f[\hat{Q}] = -\frac{\beta}{4} + \frac{\beta}{2n} \sum_{a<b}^{n} Q_{ab}^2 - \frac{1}{\beta n} \log \left[\sum_{\sigma_a} \exp \left(\beta^2 \sum_{a<b}^{n} Q_{ab} \sigma_a \sigma_b \right) \right]. \quad (1.127)$$

In the large N limit,

$$Z_n \simeq \left[\det \frac{\partial^2 f}{\partial \hat{Q}^2} \right]^{-1/2} \exp\left(-\beta n N f\left[\hat{Q}^*\right]\right), \quad (1.128)$$

where

$$\frac{\partial f}{\partial Q_{ab}} = 0, \quad (1.129)$$

defines the saddle point value of the matrix Q_{ab}.

Note that we can take the Q matrix to have zero diagonal values, thereby accounting for the fact that we have already separated the diagonal components out.

We will now go on to describing the solutions to this mean field problem. The strategy will be to parametrize the matrix in a simple manner and then to perform the saddle point evaluation.

1.9.1 Replica Symmetric Solution

It is natural to assume a symmetry between replicas of the system

$$Q_{ab} = q \quad \forall \, a \neq b \quad (1.130)$$

(It will turn out to be equivalent to the approximation that there is a unique ground state.) Thus,

$$f(q) = -\frac{1}{4}\beta + \frac{\beta}{2n} \frac{n(n-1)}{2} q^2 \quad (1.131)$$

$$- \frac{1}{\beta n} \log \left[\sum_{\sigma_a} \exp \left(\frac{\beta^2}{2} (\sum_a^n \sigma_a)^2 q - \frac{1}{2} \beta^2 n q \right) \right],$$

1.9. The Sherrington Kirkpatrick Model

where we have pulled out the diagonal terms, n of them, all of the same value. This can now be extracted from the integral as well with

$$\frac{-1}{\beta n} \times \frac{-1}{2}\beta^2 nq = \frac{\beta q}{2}. \tag{1.132}$$

Now we apply the same methods used earlier, of linearizing the quadratic expression. We use

$$\prod_{\alpha=1}^{n}\left(\sum_{\sigma_a=\pm 1}\exp\beta\sigma_a\sqrt{q}z\right) = ([2\cosh(\beta\sqrt{q}z)]^n, \tag{1.133}$$

inside

$$f(q) = -\frac{\beta}{4} + \frac{\beta q}{2} + \frac{1}{4}(n-1)\beta q^2 \tag{1.134}$$
$$-\frac{1}{\beta n}\log\left[\int_{-\infty}^{+\infty}\frac{dz}{\sqrt{\pi}}\exp\left(-\frac{1}{2}z^2\right)\prod_{\alpha=1}^{n}\left(\sum_{\sigma_a=\pm 1}\exp\beta\sigma_a\sqrt{q}z\right)\right].$$

Now writing

$$([2\cosh(\beta\sqrt{q}z)]^n = e^{n\ln[2\cosh(\beta\sqrt{q}z)]}, \tag{1.135}$$

and expanding in n, the first term is

$$1 + n\left[2\cosh(\beta\sqrt{q}z)\right] + n^2\ldots \tag{1.136}$$

The first term goes away on integration since this is a Gaussian integral normalized to 1; taking the logarithm gives zero. The second term is the important contributor and all higher order terms will vanish as $n \to 0$. Thus we may now take the limit $n \to 0$

$$f(q) = -\frac{1}{4}\beta(1-q)^2 - \frac{1}{\beta}\int_{-\infty}^{-\infty}\frac{dz}{\sqrt{2\pi}}\exp\left[-\frac{1}{2}z^2\right]\ln\left[2\cosh(\beta\sqrt{q}z)\right]. \tag{1.137}$$

We now take the derivative with respect to q to find the minimum of the replica free energy. One term we obtain is

$$\frac{1}{2}\beta(1-q). \tag{1.138}$$

The second is more complicated and looks like the integral

$$\frac{1}{2}\int_{-\infty}^{-\infty}\frac{dz}{\sqrt{2\pi q}}\exp\left[-\frac{1}{2}z^2\right]\left[z\tanh\beta\sqrt{q}z)\right]. \tag{1.139}$$

We now evaluate this through integration by parts, with $dv = \exp\left[-\frac{1}{2}z^2\right](-z)$ which implies that $v = \exp\left[-\frac{1}{2}z^2\right]$. Then $u = \left[\tanh\beta\sqrt{q}z)\right]$. Because of the exponential falloff, the uv term doesn't contribute; the only contribution is from the $\int v du$ term. This is then

$$-\frac{1}{2}\int_{-\infty}^{-\infty}\frac{dz}{\sqrt{2\pi q}}\exp\left[-\frac{1}{2}z^2\right](\beta\sqrt{q})\left[sech^2(\beta\sqrt{q}z)\right]. \quad (1.140)$$

We can now eliminate $\sqrt{q}$ from top and bottom, substitute for

$$sech^2(x) = 1 + \tanh^2(x), \quad (1.141)$$

and then get

$$\beta(1-q) = -\beta\int_{-\infty}^{-\infty}\frac{dz}{\sqrt{2\pi}}\exp\left[-\frac{1}{2}z^2\right]\left[1+\tanh^2(\beta\sqrt{q}z)\right], \quad (1.142)$$

to finally get the saddle point equation in the form

$$q = \int_{-\infty}^{+\infty}\frac{dz}{\sqrt{2\pi}}\exp\left[-\frac{1}{2}z^2\right]\tanh^2\left(\beta\sqrt{q}z\right), \quad (1.143)$$

where

$$q = \frac{1}{N}\sum_i^N\langle\sigma_i\rangle^2. \quad (1.144)$$

Note that for $T \rangle 1$ there is only the solution $q = 0$. For smaller T, there is a non-zero solution; if $(1-T) \sim \tau$, then $q(\tau) \sim \tau$. If $T \to 0$, then the solution is $q \to 1$.

The solution obtained for q is the physical order parameter

$$q = \frac{1}{N}\sum_i^N\langle\sigma_i\rangle^2. \quad (1.145)$$

The fact that q is non-zero relates to the fact that spins are frozen in a random state. Since there is only one solution for q, the ground state is unique.

This solution can be shown to have interesting properties. The specific heat and susceptibility show a cusp, as in the experiments, while the order parameter q rises smoothly at the transition point. Unfortunately, this solution is qualitatively incorrect below the transition. It has an entropy which becomes negative below T_c. This was interpreted by de Almeida and Thouless in terms of the breakdown of the stability of the replica symmetric solution. For some time it was believed that it was the replica trick which gave this erroneous result. It turns out that the answer is more subtle than that.

The error arises because the pattern of symmetry breaking we have discussed is not sophisticated enough. The correct form of the pattern of symmetry breaking was worked out by Parisi in a creative and insightful set of papers from the early 1980's. Essentially, Parisi made an ansatz for $q_{\alpha\beta}$ which goes beyond the simplest parametrization described above. He formulated the calculation in several steps, starting with the simplest, or *one-step* replica symmetry breaking. He then went on to show how an infinite hierarchy of replica symmetry breaking could be accommodated in an ansatz, the *Parisi ansatz*. This calculation, however, is complex enough that it will not be described here.

1.10 Physics of the Spin Glass State

In the replica-symmetric theory, a single parameter, the EA order parameter q, became non-zero at the transition point and could thus be taken to be the physical order parameter. However, as shown by Parisi and his collaborators, replica symmetry is broken and there is an exponentially large number of metastable states which appear just below the transition. The system could be trapped in any one of these states. Clearly, the precise nature of ordering is different in each of these states and we need an order parameter description capable of describing the essential features of these metastable states, just as the magnetization provides a way of differentiating between different ordered states in the pure ferromagnet.

To understand the physics, it is useful to introduce the notion of pure states. Consider a ferromagnet. At low temperatures, in the ordered state, the system breaks symmetry to choose one of a large number of possible ordered states. The barriers between these symmetry broken states is exponentially large in the system size. The observable state in this limit is then not the *Gibbs* state, obtained by summing over all spin configurations as in the conventional definition of the partition function, but corresponds only to those configurations which are appropriate to the particular symmetry broken state which is picked out.

Such a state is called a *pure state* and has the property that connected correlation functions, defined below, vanish at large distances:

$$\langle \sigma_i \sigma_j \rangle_c \equiv \langle \sigma_i \sigma_j \rangle - \langle \sigma_i \rangle \langle \sigma_j \rangle \tag{1.146}$$

Note that the pure states in non-disordered models are typically related to each other by symmetry, as in the case of the simple ferromagnetic Ising model.

The structure of (replica-) symmetry breaking in the spin glass state suggests that there may be a very large number of pure states in the thermodynamic limit, corresponding to the exponentially large number of states into which the system can freeze. The Gibbs state can be constructed if we knew the weight factor for each of the pure states, corresponding to the relative probability of finding them in a fixed disorder background

$$\langle \sigma_i \rangle = m_i = \sum_\alpha w_\alpha m_i^\alpha. \tag{1.147}$$

Here the statistical weights w_α can be written as

$$w_\alpha = \exp(-F_\alpha), \tag{1.148}$$

with F_α the free energy corresponding to state α. A similar relation holds for the correlation functions.

$$\langle \sigma_1 \sigma_2 \rangle = \sum_\alpha w_\alpha \langle \sigma_1 \sigma_2 \rangle_\alpha. \tag{1.149}$$

To understand how pure states in the spin glass problem might be distinguished from each other, define a quantity measuring the overlap between the two states, defined through

$$q_{\alpha\beta} = \frac{1}{N} m_i^\alpha m_i^\beta. \tag{1.150}$$

To describe the statistics of overlaps, we introduce

$$P_J(q) = \sum_{\alpha\beta} w_\alpha w_\beta \delta(q_{\alpha\beta} - q). \tag{1.151}$$

Note that this depends on the disorder. Once we disorder average we get

$$P(q) = \bar{P}_J(q). \tag{1.152}$$

The function $P(q)$ gives the probability of finding two pure states with a given overlap q, conditioned by the fact that these states are weighted by the probability of their appearance in the ensemble.

In a ferromagnet, $P(q)$ is a δ-function peak at $q = 0$ in the high temperature phase and consists of two δ-function peaks at $q = \pm m^2$. In a spin glass with a large number of pure states, unrelated by symmetry at low temperatures, one expects $P(q)$ to have far more structure. It is this function $P(q)$ which turns out to be the physical order parameter. If a disordered system exhibited only two frozen ground states related by symmetry, as in some competing theories of the spin glass transition in finite-dimensional systems, then $P(q)$ would resemble that in the pure ferromagnet.

The picture which emerges of the spin glass state order parameter from a detailed calculation due to Parisi and collaborators, is the following:

1. Just below T_c the system can be in a large number of pure states, defined as valleys constituting minima of the free energy. The configurations of the m's are different in each of these states.

2. The value of the self-overlaps $q(T) = \sum_i^N m_i^2$ appears to be the same in all the states

3. The mutual pair overlaps appear to continuously fill the interval $0 \leq q^{\alpha\beta} \leq q(T)$. The distribution of values of $q^{\alpha\beta}$ is described by a probability function $P(q)$

4. If the temperature is decreased slightly, each pure state is divided into numerous new ones. These states have a larger value of the overlap.

5. On a further decrease of the temperature, each pure state is further subdivided. This branching goes on to zero temperature.

6. In a whole temperature interval $T \leq T_c$, there is a continuous sequence of phase transitions.

It is clear that this extremely complex pattern of symmetry breaking has no parallel in the conventional phase transitions describable by Landau theory.

1.11 Modern Developments

It is now generally believed that Parisi's ansatz for replica symmetry breaking in the Sherrington-Kirkpatrick model is exact. Mathematical results from the French mathematician M. Talagrand and others prove the validity of the expression for the free energy obtained in Parisi's detailed calculation. The relevance of the Parisi ansatz to real spin glasses is still a largely open problem. The Sherrington-Kirkpatrick model is, after all, an infinite range model in which all spins are coupled to each other and thus very far from real experimental systems.

From simulations on finite-ranged Ising spin glasses, due principally to Young, Marinari and their collaborators, it appears very likely that there is a spin-glass transition in the three-dimensional nearest neighbour Ising spin glass with Gaussian disorder. There also appears to be a finite temperature transition in the Heisenberg spin glass in three dimensions.

To address the issue of spin glass ordering in finite dimensional systems, Bray and Moore and Fisher and Huse came up with an alternative view of the spin glass phase. In this picture, there is a single configuration, upto a global flip of all spins, which is the true ground state. There is no net magnetization i.e.

$$m = \frac{1}{N}\sum \langle \sigma_i \rangle = 0, \qquad (1.153)$$

while the Edwards-Anderson order parameter

$$q = \frac{1}{N}\sum_i \langle \sigma_i \rangle^2 \qquad (1.154)$$

is non-zero, signalling the freezing of spins into the disordered ground state. This view of the spin glass phase is thus very much like the conventional picture of transitions in a pure system. The only difference is in the nature of excitations out of this ground state. Since the system is disordered however, one should expect that excitations out of this ground state should be non-trivial.

Such excitations have been argued to be "**droplets**". These excitations, obtained by flipping a group of spins across a scale L, are proposed to have a typical excitation energy (or gap) which scales as

$$\Delta E_t(L) \propto L^\theta. \tag{1.155}$$

The exponent θ must be bounded as $0 < \theta < (d-1)/2$.

The kinetics of spin flips is accommodated by assuming that the typical barriers for flipping a region of size L, $B_t(L)$ scale as

$$B(L) \sim L^\Psi. \tag{1.156}$$

Within replica symmetry breaking theory $\theta = 0$ for some excitations, whereas the droplet picture has $\theta > 0$ always.

The probability distribution of energy gaps $\rho(\Delta E)$ is assumed to have a scaling form

$$\rho(\Delta E(L)) = \frac{1}{\Delta E_t(L)} \tilde{\rho}(\frac{\Delta E}{\Delta E_t(L)}). \tag{1.157}$$

Provided $\tilde{\rho}(x)$ is nonzero as $x \to 0$, there is weight for droplet excitations of arbitrarily low energy, permitting slow dynamics even for T close to zero. It is these excitations which can give the spin glass phase its characteristic complex relaxation.

This picture differs in several important ways from that proposed by Parisi. Parisi's solution indicates that a line, called the de Almeida-Thouless line, in the magnetic field-temperature plane, exists across which replica symmetry breaking occurs. This is a non-trivial attribute because the line separates an ergodic phase from a non-ergodic phase. Since the field breaks the up-down symmetry of the Hamiltonian, one has a transition without a change in symmetry. Also, the overlap distribution function of Parisi is a non-trivial one. In the droplet picture, neither of these are true. There is no AT line and the order parameter is a number as opposed to a function.

An alternative scenario which combines some of the ideas of the droplet model with intuition from the Parisi approach is the **trivial-non-trivial** scenario for finite dimensional spin glasses. The TNT scenario suggests that there are different local (at scales much less than the system size) and global exponents (for scales comparable to the system size) for the excitation energy as a function of the excitation length scale.

The local exponent is given by domain wall estimates. The global exponent is possibly consistent with $\theta_g = 0$. In the TNT scenario, one has the coexistence of a droplet model at finite length scales with mean-field behaviour for system-scale excitations. The excitations are argued to be topologically non-trivial, reaching out to the boundaries of the lattice. Physically, this is a situation where excitations look random beyond the scale of a few lattice spacings, have a surface to volume ratio of order unity and are space filling and space spanning.

1.12 K-SAT and the Spin Glass Problem

In theoretical computer science, a recurring theme is the understanding of the "computational complexity" of a given computational task. Many such tasks are surprisingly difficult to solve and are found in contexts ranging from optimization and hardware design to computational biology. A substantial class of such problems belong to what are called NP-complete problems. This class of computational tasks consists of problems where a potential solution can be checked for correctness in a time which is polynomial in the size of the input. However, *finding* such a solution *a priori*, in the worst case, takes an exponential time in the size of the input.

A large number of problems have been shown to be NP-complete. It is believed that if an efficient method for finding the solution of any one such problem is found, one would have an efficient algorithm for all NP-complete problems belonging to the same class. A fundamental conjecture of complexity theory is that no such efficient algorithm exists.

One example of an interesting class of such problems is the K-satisfiability problem, or **K-SAT**. It is stated as follows: Consider a set of N boolean variables x_i, with $i = 1 \ldots N$ and $\{x_i = 0, 1\}$. A *literal* is either a variable x_i or its negation $\neg x_i$. A *clause* C is the logical OR ($\vee$) between K distinct literals. It is thus true as soon as one of the literals is true. A formula is the logical AND ($\wedge$) between M clauses and is thus true if and only if all the clauses are true. A formula is said to be **satisfiable** if there is an assignment of the variables such that the formula is true.

To construct the K-SAT problem in a form closest to the spin glass problem, choose K among the N possible indices at random and, for each of them, choose the corresponding x_i or its negation with equal probability. Repeat this process to obtain M independently chosen clauses $\{C_\ell\}_{\ell=1,\ldots,M}$. Ask for all these clauses to be true at the same time, *i.e.* evaluating F to be true. A logical assignment of the $\{x_i\}$'s which returns F to be true is called a solution of the $K-$satisfiability problem. If no such assignment exists, F is said to be unsatisfiable. An example of a 2-SAT (here $K = 2$) formula is

$$(x \vee y) \wedge ((\neg x) \vee (\neg y)). \tag{1.158}$$

This formula can be satisfied by setting x to true and y to false. In the general case, the goal is to find out whether an assignment exists. In the worst case, a search through the space of truth assignments is required to determine that no such assignment exists or to find a suitable assignment.

A large number of problems can be shown to be instances of K-SAT. The K-SAT problem is a constraint satisfaction problem. It involves several variables, each taking values in a small domain, and some constraints, each forbidding some of the possible joint values for the variables. For a given instance, the question is whether there are values for the variables that simultaneously satisfy all the constraints. $K = 2$ belongs to the class P of polynomial problems, which can be solved by algorithms with a running time inreasing polynomially with the number of relevant variables. For $K \geq 3$, K-SAT belongs to the NP-complete class.

The spin-glass version of the K-SAT problem is related to the random version of K-SAT described above. The thermodynamic limit is achieved in the limit in which $M = \alpha N$, with both M and N going to infinity. The link with spin glasses comes from considering a group (pair, triplet etc.) of interacting spins as a constraint. We can assign the spin $\sigma_i = 1$ if the Boolean variable x_i is true and $\sigma_i = -1$ if it is false. The connection to statistical mechanics comes from assigning an energy to the system. A satisfactory assignment can be assigned zero energy, with the energy increasing with the number of unsatisfied clauses. Finding the state of minimal energy amounts to minimizing the number of violated constraints. The general difficulty both with the spin glass and the K-SAT problem arises from frustration. This makes it diffcult to find the global minimum by local flips. The case $K = 2$ then resembles the Sherrington-Kirkpatrick model. $K = 3$ and higher resemble the so-called p-spin models.

Consider formulas with a fixed ratio of clauses to variables. These exhibit the following remarkable behaviour. When the ratio is small, formulas have many variables and few constraints. There are thus many assignments which are satisfying. When the ratio is large, the variables are highly constrained and formulas almost certainly have no satisfying assignments.

Surprisingly, as the ratio $\alpha(K)$ of clauses to variables grows, the transition from probably satisfiable to probably unsatisfiable is not gradual but abrupt. This change resembles a phase transition. Random K-SAT problems appear to be hardest to solve close to this transition point. Inspired by ideas from spin glasses, Mezard, Parisi and collaborators have suggested that the hardness of random $K-$SAT is connected to the fact that for a given formula, the geometry of the space of solutions undergoes a dramatic transformation at a ratio well below the satisfiability threshold. It is suggested that it is this transition that is likely to be the key to understanding the hardness of this problem.

References

[1] S. F. Edwards and P. W. Anderson, J. Phys. F **5**, 965 (1975). This is the paper which introduces the replica trick for the first time, defining an appropriate simplified Hamiltonian for the spin glass problem.

[2] D. Sherrington and S. Kirkpatrick, Phys. Rev. Lett. **35**, 1792 (1975). This paper provides the (replica-symmetric) mean-field solution of the infinite-ranged model which bears the name of the authors.

[3] K. Binder and A. P. Young, Rev. Mod. Phys. **58**, 801 (1986). An influential and useful review.

[4] M. Mezard, G. Parisi, M.A. Virasoro, *Spin Glass theory and Beyond*, World Scientific, Singapore (1987). A book with introductory material followed by a set of reprints. Compact and rich in ideas, but requires research-level training.

[5] S. Kirkpatrick, C. D. Gelatt, Jr. and M. P. Vecchi, Science, **220**, 671 (1983). The use of physics-inspired ideas, the concept of annealing to find the ground state, in providing heuristic solutions to computationally hard problems, such as the travelling salesman problem.

[6] *Spin Glasses and Random Fields*, A.P. Young, ed., World Scientific, Singapore, (1998). An highly useful set of review articles covering various aspects of these fields at a more technical level.

[7] A. Houdayer and O.C. Martin, Europhys. Lett **49(6)**, 794 (2000); M. Palassini, and A. P. Young, Phys. Rev. Lett. **83**, 5216 (1999); F. Krzakala and O.C. Martin Phys. Rev. Lett. **85**, 3013 (2000). The trivial-non-trivial (TNT) scenario for the dimension dependence of spin glass physics.

[8] P. Nordblad, L. Lundgren and L. Sandlund, J. Mag. and Mag. Mater. **54**, 185 (1986). A classic reference for the experiments.

[9] V. S. Dotsenko, *Introduction to the Theory of Spin Glasses and Neural Networks*, World Scientific, Singapore (1995). A very useful text which works everything out, building up from basics. Unfortunately, the reader should beware of a large number of misprints.

[10] D. Stauffer, Phys. Rep. **54**, 1 (1979). A review of percolation theory.

[11] D. Stauffer and A. Aharony, *Introduction to Percolation Theory* 2nd ed., Taylor and Francis, London (1994). A useful and comprehensive book on percolation theory.

[12] M. Sahimi, *Applications of Percolation Theory*, Taylor and Francis, London (1994)

[13] D.S. Fisher and D.A Huse, Phys. Rev. Lett **56**, 1601 (1986)

[14] M. Mezard, G. Parisi and R. Zecchina, Science **297**, 812 (2002). The study of K-SAT problems through statistical physics methods inspired by spin glass physics. Clear and well-written.

[15] J. S Yedidia, Lectures at the Santa Fe Summer School on Complex Systems 1992. Available online at *http://nerdwisdom.files.wordpress.com/2007/08/santafe.pdf*. A very useful and pedagogical treatment of several issues, including variational and cavity methods.

Chapter 2

Phase Transitions in Disordered Quantum Systems: Transverse Ising Models

Bikas K. Chakrabarti and Arnab Das

We introduce the transverse Ising model as a prototype for discussing quantum phase transitions. Mean field theory and its application to superconductivity by BCS are discussed, as well as real space renormalization group techniques in one dimension. Next, we introduce the Suzuki-Trotter formalism to show the correspondence between d-dimensional quantum systems and $(d+1)$-dimensional classical systems. We then discuss transverse Ising spin glass models, namely the Sherrington-Kirkpatrick (SK) model, the Edwards-Anderson (EA) model and the $\pm J$ model. We briefly discuss the mean field, exact diagonalization and quantum Monte Carlo results for their phase diagrams. We discuss the question of replica symmetry restoration in quantum spin glasses due to the possibility of tunneling through the barriers. Finally, we discuss the quantum annealing technique and indicate its relationship with replica symmetry restoration in quantum spin glasses.

In many physical systems, cooperative interactions between spin-like (two-state) degrees of freedom tend to establish some kind of order in the system, while the presence of a non-cooperative effect (like temperature, external transverse field, tunneling etc.) tends to destroy it. The transverse Ising model can be successfully employed to study the order-disorder transitions in many such systems.

An example of the above is the study of ferro-electric ordering in Potassium Di-hydrogen Phosphate (KDP) type systems (see, e.g., [1]). To understand such ordering, the basic structure can be viewed as a lattice, where at each lattice point there is a double-well potential created by an oxygen atom and the

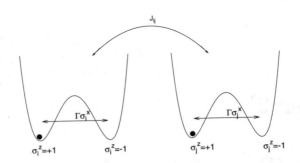

Figure 2.1: The double wells at each site (e.g., provided by oxygen in KDP) provide two (low-lying) states of the proton (shown by each double well) indicated by the Ising states $|\uparrow\rangle$ and $|\downarrow\rangle$ at each site. The tunneling between the states are induced by the transverse field term ($\Gamma\sigma^x$). The dipole-dipole interaction J_{ij} here for the (asymmetric) choice of one or the other well at each site induces the *exchange* interaction as shown.

proton resides within it in any of the two wells. In the corresponding Ising (or pseudo-spin) picture, the state of a double-well with a proton at the left-well or at the right-well are represented by $|\uparrow\rangle$ and $|\downarrow\rangle$ respectively (for a portion of the lattice see Fig. 2.1).

The protons at neighbouring sites have a mutual dipolar repulsion. Had the proton been a classical particle, the zero-temperature configuration of the system would have been one with either all the protons residing in their respective left-wells or with all residing in the right wells, corresponding to the all-up or all-down configuration of the spin system in the presence of cooperative interactions alone and at zero-temperature. Hence, since there are no thermal fluctuations at zero temperature, the Hamiltonian for the system in the corresponding pseudo-spin picture would be identical to an Ising Hamiltonian without a transverse term. However, since the proton is a quantum particle, there is always a finite probability for it to tunnel through the finite barrier between two wells even at zero-temperature *i.e.* quantum fluctuations are present even at zero temperature.

To formulate the term for the tunneling in the corresponding spin-picture, we notice that σ^x is the right operator to represent the tunneling, since one can easily see, by simple matrix multiplication, that

$$\sigma^x|\uparrow\rangle = |\downarrow\rangle \quad \text{and} \quad \sigma^x|\downarrow\rangle = |\uparrow\rangle, \tag{2.1}$$

where $|\uparrow\rangle$ represents the state where the proton is in the left well while $|\downarrow\rangle$ represents the state with the proton in the right well. The tunneling term will then be exactly represented by the transverse field term in the transverse Using Hamiltonian. Here the transverse field coefficient Γ will represent the tunneling integral, which depends on the width & height of the barrier, the mass of the particle etc.

2.1 Transverse Ising Model

The system discussed above can be represented by a quantum Ising system, having Hamiltonian

$$\mathcal{H} = -\sum_{\langle i,j \rangle} J_{ij} \sigma_i^z \sigma_j^z - \Gamma \sum_i \sigma_i^x. \tag{2.2}$$

Here J_{ij} is the coupling between the spins at sites i and j, and σ^α's ($\alpha = x, y, z$) are Pauli spins satisfying the commutation relations

$$[\sigma_i^\alpha, \sigma_j^\beta] = 2i\delta_{ij}\epsilon_{\alpha\beta\gamma}\sigma_i^\gamma \tag{2.3}$$

Also, δ_{ij} is the Kronecker δ, while $\epsilon_{\alpha\beta\gamma}$ is the Levi-Civita symbol and $\langle i,j \rangle$ in (2.2) represents neighbouring pairs.

The Pauli spin matrices are representatives of spin-1/2. This implies that σ^z has two eigenvalues (± 1), corresponding to spins aligned either along z-direction or along the opposite direction respectively. The eigenstate corresponding to eigenvalue ($+1$) is symbolically denoted by $|\uparrow\rangle$, while that corresponding to (-1) is denoted by $|\downarrow\rangle$.

If we represent

$$|\uparrow\rangle \Leftrightarrow \begin{pmatrix} 1 \\ 0 \end{pmatrix} \tag{2.4}$$

and

$$|\downarrow\rangle \Leftrightarrow \begin{pmatrix} 0 \\ 1 \end{pmatrix}, \tag{2.5}$$

then taking these two eigen-vectors as basis, the Pauli spins have the following matrix representations

$$\sigma^x = \begin{pmatrix} 0 & 1 \\ 1 & 0 \end{pmatrix}, \quad \sigma^y = \begin{pmatrix} 0 & -i \\ i & 0 \end{pmatrix}, \quad \sigma^z = \begin{pmatrix} 1 & 0 \\ 0 & -1 \end{pmatrix}. \tag{2.6}$$

With these, one can see that the relations in (2.3) are easily satisfied and the tunneling required in (2.1) can be easily accommodated. The order parameter for such a system is generally taken to be the expectation value of the z-component of the spin, i.e. $\langle \sigma^z \rangle$. In such a system, absolute ordering with complete alignment along the z-direction is not possible even at zero-temperature, i.e., $\langle \sigma^z \rangle_{T=0} \neq 1$, when $\Gamma \neq 0$. In general, therefore, the order ($\langle \sigma^z \rangle \neq 0$) to disorder $\langle \sigma^z \rangle = 0$ transition can be brought about by tuning either or both of the tunneling field Γ and the temperature T (see Fig. 2.2).

2.2 Mean Field Theory

2.2.1 For T = 0 [2]

Let

$$\sigma_i^z = |\vec{\sigma}|\cos\theta, \quad \text{and} \quad \sigma_i^x = |\vec{\sigma}|\sin\theta, \tag{2.7}$$

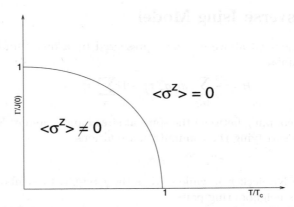

Figure 2.2: Schematic phase diagram of the model represented by Hamiltonian (2.2).

where θ is the angle between $\vec{\sigma}$ and z-axis. This renders the two mutually non-commuting parts of the Hamiltonian Eq. (2.2) commuting, since both are expressed in terms of the $|\vec{\sigma}|$ operator only. If σ is the eigen-value of $|\vec{\sigma}|$ ($\sigma = 1$ for Pauli spins), then the energy per site of the semi-classical system is given by

$$E = -\sigma \Gamma \sin\theta - \sigma^2 J(0) \cos^2\theta, \qquad (2.8)$$

$J(0) = J_i(0) = \sum_{\langle i,j \rangle} J_{ij}$, where j indicates the j-th nearest neighbour of the i-th site.

The average of the spin-components is given by

$$\langle \sigma^z \rangle = \cos\theta \qquad (2.9)$$
$$\langle \sigma^x \rangle = \sin\theta. \qquad (2.10)$$

The energy (2.8) is minimized for

$$\sin\theta = \Gamma/J(0) \qquad \text{or,} \qquad \cos\theta = 0. \qquad (2.11)$$

Thus we see that if $\Gamma = 0$, $\langle \sigma^x \rangle = 0$ and the order parameter $\langle \sigma^z \rangle = 1$, indicating perfect order. On the other hand, if $\Gamma < J(0)$, then the ground state is partially polarized, since neither of $\langle \sigma^z \rangle$ or $\langle \sigma^x \rangle$ is zero. However, if $\Gamma \geq J(0)$, we must have $\cos\theta = 0$ for the ground state energy, which means $\langle \sigma^z \rangle = 0$, i.e., the state is a completely disordered one. Thus, as Γ increases from 0 to J(0), the system undergoes a transition from an ordered (ferro)- phase with order parameter $\langle \sigma^z \rangle = 1$ to disordered (para)-phase with order parameter $\langle \sigma^z \rangle = 0$ (see Fig. 2.2).

2.2.2 For T $\neq$ 0 [3, 4]

The mean field method can also be extended to obtain the behaviour of this model at non-zero temperature. In this case we define a mean field $\tilde{\mathbf{h}}_i$ at each

2.2. Mean Field Theory

site i, which is, in a sense, a resultant of the average cooperative reinforcement in the z-direction and the applied transverse field in x-direction. Specifically, we take, for the general random case,

$$\tilde{\mathbf{h}}_i = \Gamma \hat{x} + \left(\frac{1}{2}\sum_j J_{ij}\langle \sigma_j^z \rangle\right)\hat{z}, \quad (2.12)$$

and the spin-vector at the i-th site follows $\tilde{\mathbf{h}}_i$. The spin-vector at the i-th site is given by

$$\vec{\sigma}_i = \sigma_i^x \hat{x} + \sigma_i^z \hat{z}, \quad (2.13)$$

and the Hamiltonian thus reads

$$\mathcal{H} = -\sum_i \tilde{\mathbf{h}}_i \cdot \vec{\sigma}_i. \quad (2.14)$$

For the non-random case, all the sites have identical environments, hence $\tilde{\mathbf{h}}_i$ is replaced by $\tilde{\mathbf{h}} = \Gamma \hat{x} + \langle \sigma^z \rangle J(0)$. The resulting Hamiltonian takes the form

$$\mathcal{H} = -\tilde{\mathbf{h}} \cdot \sum_i \vec{\sigma}_i. \quad (2.15)$$

The spontaneous magnetization can readily be written down as

$$\tilde{\sigma} = \tanh(\beta|\tilde{\mathbf{h}}|) \cdot \frac{\tilde{\mathbf{h}}}{|\tilde{\mathbf{h}}|} \quad (2.16)$$

$$|\tilde{\mathbf{h}}| = \sqrt{\Gamma^2 + (J(0)\langle \sigma^z \rangle)^2}. \quad (2.17)$$

Now if $\tilde{\mathbf{h}}$ makes an angle θ with z-axis, then $\cos\theta = J(0)\langle \sigma^z \rangle/|\tilde{\mathbf{h}}|$ and $\sin\theta = \Gamma/|\tilde{\mathbf{h}}|$. Hence we have

$$\langle \sigma^z \rangle = |\tilde{\mathbf{h}}|\cos\theta = [\tanh(\beta|\tilde{\mathbf{h}}|)]\left(\frac{J(0)\langle \sigma^z \rangle}{|\tilde{\mathbf{h}}|}\right), \quad (2.18)$$

and

$$\langle \sigma^x \rangle = [\tanh(\beta|\tilde{\mathbf{h}}|)]\frac{\Gamma}{|\tilde{\mathbf{h}}|}. \quad (2.19)$$

Here, $\beta = (1/k_B T)$.

Equation (2.19) is the self-consistency equation which can be solved graphically or otherwise to obtain the order parameter $\langle \sigma^z \rangle$ at any temperature T and transverse field Γ. Clearly, the order-disorder transition is tuned both by Γ and T (see Fig. 2.2).

2.2.3 For $\Gamma = 0$ (Transition driven by T):

$$\langle \sigma^z \rangle = \tanh\left(\frac{J(0)\langle \sigma^z \rangle}{k_B T}\right) \qquad (2.20)$$

and

$$\langle \sigma^x \rangle = 0 \qquad (2.21)$$

One can easily see graphically that the above equations have a nontrivial solution only if $k_B T < J(0)$, i.e. $\langle \sigma^z \rangle \neq 0$ for $k_B T < J(0)$: $\langle \sigma^z \rangle = 0$ for $k_B T > J(0)$. This shows that there is a critical temperature $T_c = J(0)$ above which there is no order.

2.2.4 For $k_B T \to 0$ (Transition driven by Γ):

$$\langle \sigma^z \rangle = \frac{J(0)\langle \sigma^z \rangle}{\sqrt{(\Gamma)^2 + (J(0)\langle \sigma^z \rangle)^2}} \qquad \left(\text{since,} \quad \tanh x \Big|_{x \to \infty} = 1\right).$$

From this equation we easily see that in the limit $\frac{\Gamma}{J(0)} \to 1$, the only real nontrivial solution is

$$\langle \sigma^z \rangle \to 0 \qquad (2.22)$$

and

$$\langle \sigma^x \rangle = \frac{\Gamma}{\sqrt{(\Gamma)^2 + (J(0)\langle \sigma^z \rangle)^2}} \to 1, \qquad \text{as} \quad \frac{\Gamma}{J(0)} \to 1. \qquad (2.23)$$

Thus, there is a critical transverse field $\Gamma_c = J(0)$ such that for any $\Gamma > \Gamma_c$ there is no order even at zero temperature. In general one observes that at any temperature $T < T_c$, there exists some transverse field Γ_c at which the transition from the ordered state ($\langle \sigma^z \rangle \neq 0$) to the disordered state ($\langle \sigma^z \rangle = 0$) occurs. The equation for the phase boundary in the ($\Gamma - T$) - plane is obtained by putting $\langle \sigma^z \rangle \to 0$ in equation (2.19). The equation gives the relation between Γ_c and T_c as follows

$$\tanh\left(\frac{\Gamma_c}{k_B T}\right) = \frac{\Gamma_c}{J(0)}. \qquad (2.24)$$

One may note that for ordered phase, since $\langle \sigma^z \rangle \neq 0$,

$$\frac{1}{|\tilde{\mathbf{h}}|} \tanh(\beta|\tilde{\mathbf{h}}|) = \frac{1}{J(0)} = \text{Constant}. \qquad (2.25)$$

Hence, $\langle \sigma^x \rangle = \frac{\Gamma}{|\tilde{\mathbf{h}}|} \tanh(\beta|\tilde{\mathbf{h}}|) = \frac{\Gamma}{J(0)}$; independent of temperature in the ordered phase. While for the disordered phase, since $\langle \sigma^z \rangle = 0$,

$$\langle \sigma^x \rangle = \tanh(\beta \Gamma). \qquad (2.26)$$

2.2.5 Dynamic Mode-softening Picture

The elementary excitations in a system as described above are known as spin waves. They can be studied by employing the Heisenberg equation of motion for σ^z using the Hamiltonian. The equation of motion is then given by

$$\dot{\sigma}_i^z = (i\hbar)^{-1}[\sigma_i^z, \mathcal{H}] \tag{2.27}$$

or,

$$\dot{\sigma}_i^z = 2\Gamma \sigma_i^y \quad \text{(with} \quad \hbar = 1) \tag{2.28}$$

Hence,

$$\ddot{\sigma}_i^z = 2\Gamma \dot{\sigma}_i^y = 4\Gamma \sum_j J_{ij}\sigma_i^z\sigma_i^x - 4\Gamma^2 \sigma_i^z \tag{2.29}$$

Using Fourier transforms and the random phase approximation ($\sigma_i^x \sigma_j^z = \sigma_i^x \langle \sigma_j^z \rangle + \langle \sigma_i^x \rangle \sigma_j^z$, with $\langle \sigma^z \rangle = 0$ in para phase), we get

$$\omega_q^2 = 4\Gamma(\Gamma - J(q)\langle \sigma^x \rangle), \tag{2.30}$$

for the elementary excitations (where $J(q)$ is the Fourier transform of J_{ij}). The mode corresponding to ($q = 0$) softens, *i.e.*, ω_0 vanishes at the same phase boundary given by Eq. (2.24).

2.3 BCS Theory of Superconductivity

The phonon mediated effective attractive interaction between electrons give rise to a cooperative quantum Hamiltonian. Although the quantum phase transition in such a system is not physical or meaningful, the finite temperature superconducting phase transition can be studied easily following the mean field theory discussed here (using a pseudo-spin mapping [5]). The relevant part of the Hamiltonian of electrons that take part in superconductivity has the following form

$$\mathcal{H} = \sum_k \epsilon_k (c_k^\dagger c_k + c_{-k}^\dagger c_{-k}) - V \sum_{kk'} c_{k'}^\dagger c_{-k'}^\dagger c_{-k} c_k \tag{2.31}$$

Here the suffix k indicates a state with momentum $\vec{k}$ and spin up, while $(-k)$ indicates a state with momentum $-\vec{k}$ and spin down and V is a positive constant that models the attractive coupling between electrons through phonons. We will solve this equation following spin-analog method, see Ref. [4]. Here we are considering low-lying states containing a pair of electrons $(k, -k)$. For a given k, there are two possible states that come into consideration: either the pair exists, or it does not. Thus we enter into a spin-like two-state picture as follows.

Let us introduce the number operator $\hat{n}_k = c_k^\dagger c_k$. This reduces the Hamiltonian (2.31) to

$$\mathcal{H}_{red} = -\sum_k \epsilon_k (1 - \hat{n}_k - \hat{n}_{-k}) - V \sum_{kk'} c_{k'}^\dagger c_{-k'}^\dagger c_{-k} c_k. \qquad (2.32)$$

Here we have introduced a term $-\sum_k \epsilon_k$ with the choice $\sum_k \epsilon_k = 0$ in mind, for all k's (basically these sums are over the states within energy $\pm \omega_D$ about the Fermi level, where ω_D is the Debye energy) that participate in pair formation. As stated earlier, if n_k denotes the number of electrons in k-state, then we are considering only a subspace of states defined by $n_k = n_{-k}$, where either both of the states in the pair $(k,-k)$ are occupied, or both are empty. Now if we denote by $|1_k 1_{-k}\rangle$ a $(k,-k)$ pair-occupied state and by $|0_k 0_{-k}\rangle$ an unoccupied one, then

$$(1 - \hat{n}_k - \hat{n}_{-k})|1_k 1_{-k}\rangle = (1 - 1 - 1)|1_k 1_{-k}\rangle = -|1_k 1_{-k}\rangle, \qquad (2.33)$$

and

$$(1 - \hat{n}_k - \hat{n}_{-k})|0_k 0_{-k}\rangle = (1 - 0 - 0)|0_k 0_{-k}\rangle = |0_k 0_{-k}\rangle \qquad (2.34)$$

Thus we switch over to our familiar pseudo-spin picture through the following correspondence

$$|1_k 1_{-k}\rangle \Leftrightarrow |\downarrow\rangle_k, \qquad (2.35)$$

$$|0_k 0_{-k}\rangle \Leftrightarrow |\uparrow\rangle_k, \qquad (2.36)$$

$$\text{and} \quad (1 - n_k - n_{-k}) \Leftrightarrow \sigma_k^z. \qquad (2.37)$$

Since

$$c_k^\dagger c_{-k}^\dagger |\uparrow\rangle_k = |\downarrow\rangle_k, \quad c_k^\dagger c_{-k}^\dagger |\downarrow\rangle_k = 0, \quad \text{while} \quad c_{-k} c_k |\downarrow\rangle_k = |\uparrow\rangle_k \qquad (2.38)$$
$$\text{and} \quad c_{-k} c_k |\uparrow\rangle_k = 0,$$

we immediately identify its correspondence with the raising and lowering operators σ^+/σ^-:

$$\sigma^- = \sigma^x - i\sigma^y = \begin{pmatrix} 0 & 0 \\ 2 & 0 \end{pmatrix}$$

and

$$\sigma^+ = \begin{pmatrix} 0 & 2 \\ 0 & 0 \end{pmatrix}$$

and therefore

$$c_k^\dagger c_{-k}^\dagger = \frac{1}{2}\sigma_k^-, \quad c_{-k} c_k = \frac{1}{2}\sigma_k^+. \qquad (2.39)$$

2.3. BCS Theory of Superconductivity

Hence in terms of these spin operators, the Hamiltonian (2.32) takes the form

$$\mathcal{H} = -\sum_k \epsilon_k \sigma_k^z - \frac{1}{4} V \sum_{kk'} \sigma_{k'}^- \sigma_k^+. \tag{2.40}$$

Since the term $\sum_{kk'}(\sigma_{k'}^x \sigma_k^y - \sigma_{k'}^y \sigma_k^x)$ vanishes due to symmetric summing done over k and k', the Hamiltonian finally reduces to

$$\mathcal{H} = -\sum_k \epsilon_k \sigma_k^z - \frac{1}{4} V \sum_{kk'} (\sigma_{k'}^x \sigma_k^x + \sigma_{k'}^y \sigma_k^y). \tag{2.41}$$

To obtain the energy spectrum of the pseudo-spin BCS Hamiltonian (2.41) we apply now the mean field theory developed in an earlier section.

Weiss' Mean Field Solution

Just as we did in the case of the transverse Ising Hamiltonian, we introduce an average effective field $\tilde{\mathbf{h}}_k$ for each pseudo-spin σ_k as

$$\tilde{\mathbf{h}}_k = \epsilon_k \hat{z} + \frac{1}{2} V \sum_{k'} (\langle \sigma_{k'}^x \rangle \hat{x} + \langle \sigma_{k'}^y \rangle \hat{y}). \tag{2.42}$$

Consequently the Hamiltonian (2.41) takes the form

$$\mathcal{H} = -\sum_k \tilde{\mathbf{h}}_k \cdot \tilde{\sigma}_k. \tag{2.43}$$

Here for each k there is an independent spin $\tilde{\sigma}_k$ which interacts only with some effective field $\tilde{\mathbf{h}}_k$, and our system is a collection of such mutually non-interacting spins under a field $\tilde{\mathbf{h}}_k$.

Now if we redefine our x-axis along the projection of $\tilde{\mathbf{h}}_\mathbf{k}$ on the x-y plane for each k, then with all $\langle \sigma_{k'}^y \rangle = 0$ we get

$$\tan \theta_k = \frac{h_k^x}{h_k^z} = \frac{\frac{1}{2} V \sum_{k'} \langle \sigma_{k'}^x \rangle}{\epsilon_k}, \tag{2.44}$$

where θ_k is the angle between z-axis and $\tilde{\mathbf{h}}_k$.

Excitation Spectra at T = 0

Since at $T = 0$ $\langle \sigma^x \rangle = 1$,

$$\langle \sigma_{k'}^x \rangle = |\tilde{\sigma}| \sin \theta_{k'} = \sin \theta_{k'} \tag{2.45}$$

Thus from (2.44) we get

$$\tan \theta_k = (v/2\epsilon_k) \sum_{k'} \sin \theta_{k'} \tag{2.46}$$

Now let us define
$$\Delta \equiv \frac{1}{2} V \sum_{k'} \sin \theta_{k'},$$
so that $\tan \theta_k = \Delta/\epsilon_k$. Then simple trigonometry gives
$$\sin \theta_k = \frac{\Delta}{\sqrt{\Delta^2 + \epsilon_k^2}} \quad ; \quad \cos \theta_k = \frac{\epsilon_k}{\sqrt{\Delta^2 + \epsilon_k^2}}. \quad (2.47)$$

Substituting for $\sin \theta_{k'}$ into the above equation we get
$$\Delta = \frac{1}{2} V \sum_{k'} \frac{\Delta}{\sqrt{\Delta^2 + \epsilon_{k'}^2}}. \quad (2.48)$$

Assuming the spectrum to be nearly continuous, we replace the summation by an integral and note that V is attractive for energy within $\pm \omega_D$ on both sides of Fermi level; ω_D being of the order of Debye energy. Then the last equation becomes
$$1 = \frac{1}{2} V \rho_F \int_{-\omega_D}^{\omega_D} \frac{d\epsilon}{\sqrt{\Delta^2 + \epsilon^2}} = V \rho_F \sinh^{-1}(\omega_D/\Delta). \quad (2.49)$$

Here ρ_F is the density of states at Fermi level. Thus
$$\Delta = \frac{\omega_D}{\sinh(1/V\rho_F)} \cong 2\omega_D e^{-1/V\rho_F}, \quad (\text{if} \quad \rho_F V \ll 1) \quad (2.50)$$

We see that Δ is positive if V is positive. To interpret the parameter Δ, we notice that at first approximation, the excitation spectrum is obtained as the energy $\mathcal{E}_k$ to reverse a pseudo-spin in the field $\tilde{\mathbf{h}}_k$, i.e.,
$$\mathcal{E}_k = 2|\tilde{\mathbf{h}}_k| = 2\left(\epsilon_k^2 + \Delta^2\right)^{1/2}. \quad (2.51)$$

From this expression we clearly see that the minimum excitation energy is 2Δ, i.e. Δ gives the energy gap in the excitation spectrum.

Estimating the Transition Temperature T_c

To find the critical temperature for the BCS transition, we just extend the non-zero temperature version of mean field theory done for Ising case. We should have (unlike as in (2.51), where $\langle \sigma_k \rangle = 1$) for $T = 0$):
$$\langle \sigma_k^z \rangle = \tanh \left(\beta |\tilde{\mathbf{h}}_k|\right). \quad (2.52)$$

Equation (2.44) accordingly modifies to
$$\tan \theta_k = \left(\frac{V}{2\epsilon_k}\right) \sum_{k'} \tanh \left(\beta |\tilde{\mathbf{h}}_{k'}|\right) \sin \theta_{k'} \equiv \frac{\Delta(T)}{\epsilon_k}, \quad (2.53)$$

where $\Delta(T) = \frac{V}{2}\sum_{k'} \tanh\left(\frac{|\vec{\tilde{h}}_{k'}|}{T}\right)\sin\theta_{k'}$. From equation (2.51) we have

$$|\tilde{\mathbf{h}}_k| = [\epsilon_k^2 + \Delta^2(T)]. \tag{2.54}$$

The BCS transition is characterized by the vanishing of the gap Δ, since without such a gap in the spectrum, infinite conductance would not be possible except at $T = 0$. Hence, as $T \to T_c$, $\Delta \to 0$, i.e., using (2.51), $|\tilde{\mathbf{h}}_k| = \epsilon_k$ and putting this and relations like (2.47) in (2.53), we get

$$1 = V\sum_{k'} \frac{1}{2\epsilon_{k'}} \tanh\left(\frac{\epsilon_{k'}}{T_c}\right). \tag{2.55}$$

The above relation is correct if we consider an excited pair as a single entity. However, if we extend our picture to incorporate single particles excited symmetrically in momentum space, then we double the number of possible excitations, thereby doubling the overall entropy. This is exactly equivalent to a doubling of the temperature in free energy. The energy contribution to the free energy, however, remains unaltered, since two single particle excitations of same $|\tilde{\mathbf{k}}|$ have same energy as that of a pair of equal $|\tilde{\mathbf{k}}|$. Hence we replace T_c by $2T_c$, and in the continuum limit, get

$$\frac{2}{V\rho_F} = \int_{-\omega_D}^{\omega_D} \frac{d\epsilon}{\epsilon} \tanh\left(\frac{\epsilon}{2T_c}\right) = 2\int_0^{\omega_D/2T_c} \frac{\tanh x}{x} dx, \tag{2.56}$$

with $(x = \epsilon/2T_c)$ This is the equation from which we obtain T_c on integration. If $T_c \ll \omega_D$, then we may approximate $\tanh x \approx 1$, for $x \geq 1$, and for $x \ll 1$, we set $\tanh x \approx x$. This readily reduces the integral to the value $1 + \log(\omega_D/2T_c)$, from which we have

$$T_c = (e/2)\omega_D e^{-1/V\rho_F}.$$

Graphical integration gives a closer result

$$T_c = 1.14\omega_D e^{-1/V\rho_F}. \tag{2.57}$$

Comparing Eqs. (2.50) and (2.57) we get the approximate relationship

$$2\Delta \simeq 3.5 T_c. \tag{2.58}$$

This result is quite consistent with the experimental values for a number of materials. For example, the value of $2\Delta/T_c$ are 3.5, 3.4, 4.1, 3.3 for Sn, Al, Pb, and Cd superconductors respectively.

2.4 Real Space Renormalization for the Transverse Ising Chain

Here the basic idea of real space block renormalization, as in Refs. [4,6], is illustrated by applying it to an Ising chain in transverse field. Taking the cooperative

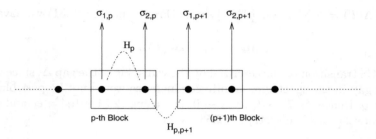

Figure 2.3: The linear chain is broken up into blocks of size b ($= 2$ here) and the Hamiltonian (2.59) can be written as the sum of block Hamiltonian's $\mathcal{H}_p$ and inter-block Hamiltonian's $\mathcal{H}_{p,p+1}$. The Hamiltonian $\mathcal{H}_p$ is diagonalized exactly and the lowest lying two states are identified as the renormalized spin states in terms of which the inter-block Hamiltonian is rewritten to get the RG recursion relation.

interaction along x-axis and the transverse field along z-axis, the Hamiltonian reads

$$\mathcal{H} = -\Gamma \sum_{i=1}^{N} \sigma_i^z - J \sum_{i=1}^{N-1} \sigma_i^x S_{i+1}^x \qquad (2.59)$$

$$= \mathcal{H}_B + \mathcal{H}_{IB}, \quad \text{(say)} \qquad (2.60)$$

Here

$$\mathcal{H}_B = \sum_{p=1}^{N/b} \mathcal{H}_p \quad ; \quad \mathcal{H}_p = -\sum_{i=1}^{b} \Gamma \sigma_{i,p}^z - \sum_{i-1}^{b-1} J \sigma_{i,p}^x \sigma_{i+1,p}^x \qquad (2.61)$$

and

$$\mathcal{H}_{IB} = \sum_{p=1}^{N/(b-1)} \mathcal{H}_{p,p+1} \quad ; \quad \mathcal{H}_{p,p+1} = -J \sigma_{b,p}^x \sigma_{1,p+1}^x. \qquad (2.62)$$

The above rearrangement of the Hamiltonian recasts the picture of N spins with nearest-neighbour interaction into one in which there are $N/(b-1)$ blocks, each consisting of a b number of spins. The part $\mathcal{H}_B$ represents the interaction between the spins within the blocks, while $\mathcal{H}_{IB}$ represents the interactions between the blocks through their terminal spins (see Fig. (2.3)).

Here we will consider $b = 2$, as shown in the figure. Now $\mathcal{H}_p$ has 4 eigenstates, and one can express them in terms of the linear superposition of the eigen-states of $\sigma_{1,p}^z \otimes \sigma_{2,p}^z$; namely,

$$|\uparrow\uparrow\rangle, \quad |\downarrow\downarrow\rangle, \quad |\uparrow\downarrow\rangle, \quad \text{and} \quad |\downarrow\uparrow\rangle. \qquad (2.63)$$

2.4. Real Space Renormalization for the Transverse Ising Chain

Considering the orthonormality of the eigen-states, one may easily see that the eigenstates of $\mathcal{H}_p$ can be expressed as

$$|0\rangle = \frac{1}{\sqrt{1+a^2}}(|\uparrow\uparrow\rangle + a|\downarrow\downarrow\rangle) \tag{2.64}$$

$$|1\rangle = \frac{1}{\sqrt{2}}(|\uparrow\downarrow\rangle + |\downarrow\uparrow\rangle) \tag{2.65}$$

$$|2\rangle = \frac{1}{\sqrt{2}}(|\uparrow\downarrow\rangle - |\downarrow\uparrow\rangle) \tag{2.66}$$

$$|3\rangle = \frac{1}{\sqrt{1+a^2}}(a|\uparrow\uparrow\rangle - |\downarrow\downarrow\rangle). \tag{2.67}$$

Here a is a coefficient required to be chosen so that $|0\rangle$ and $|3\rangle$ are eigenstates of $\mathcal{H}_p$. One gets

$$\mathcal{H}_P|0\rangle = \mathcal{H}_p\left[\frac{1}{\sqrt{1+a^2}}|\uparrow\uparrow\rangle + a|\downarrow\downarrow\rangle\right] \tag{2.68}$$

$$= [-\Gamma(\sigma_1^z + \sigma_2^z) - J(\sigma_1^x \sigma_2^x)]\frac{1}{\sqrt{1+a^2}}(|\uparrow\uparrow\rangle + a|\downarrow\downarrow\rangle) \tag{2.69}$$

$$= \frac{1}{\sqrt{1+a^2}}[-\Gamma(2|\uparrow\uparrow\rangle - 2a|\downarrow\downarrow\rangle) - J(|\downarrow\downarrow\rangle) + a|\uparrow\uparrow\rangle] \tag{2.70}$$

$$= -(2\Gamma + Ja)\frac{1}{\sqrt{1+a^2}}\left[|\uparrow\uparrow\rangle + \left(-\frac{2\Gamma - J/a}{2\Gamma + Ja}\right)a|\downarrow\downarrow\rangle\right] \tag{2.71}$$

Thus for $|0\rangle$ to be an eigenstate of $\mathcal{H}_p$, one must have

$$-\frac{2\Gamma - J/a}{2\Gamma + Ja} = 1 \tag{2.72}$$

$$\Rightarrow \quad Ja^2 - 4\Gamma a - J = 0 \tag{2.73}$$

or, $\quad a = \frac{\pm\sqrt{4\Gamma^2 + J^2} - 2\Gamma}{J}. \tag{2.74}$

To minimize the energy, we have to choose,

$$a = \frac{\sqrt{4\Gamma^2 + J^2} - 2\Gamma}{J}. \tag{2.75}$$

One can now see, applying $\mathcal{H}_p$ on its eigen-states,

$$\mathcal{H}_p|0\rangle = E_0|0\rangle, \quad E_0 = -\sqrt{4\Gamma^2 + J^2} \tag{2.76}$$
$$\mathcal{H}_p|1\rangle = E_1|1\rangle, \quad E_1 = -J \tag{2.77}$$
$$\mathcal{H}_p|2\rangle = E_2|2\rangle, \quad E_2 = +J \tag{2.78}$$
$$\mathcal{H}_p|3\rangle = E_3|3\rangle, \quad E_3 = +\sqrt{4\Gamma^2 + J^2}. \tag{2.79}$$

Now we define our new renormalized spin variables σ''s, each replacing a block in the original Hamiltonian. We retain only the two lowest lying states $|0\rangle$ and $|1\rangle$ of a block and define the corresponding $\sigma_p'^Z$ to have them as its two eigenstates, $|\uparrow\rangle = |0\rangle$ and $|\downarrow\rangle = |1\rangle$. We also define

$$\sigma'^x = \frac{\sigma_1^x \otimes \mathcal{I} + \mathcal{I} \otimes \sigma_2^x}{2}, \qquad (2.80)$$

where $\mathcal{I}$ is the 2×2 identity matrix. Now since

$$\langle 0|\sigma'^x|1\rangle = \frac{1+a}{\sqrt{2(1+a^2)}}, \qquad (2.81)$$

we take our renormalized J to be

$$J' = J\frac{(1+a)^2}{2(1+a^2)}, \qquad (2.82)$$

and since the energy gap between $|0\rangle$ and $|1\rangle$ must be equal to $2\Gamma'$ (this gap was 2Γ in the unrenormalized states), we set

$$\Gamma' = \frac{E_1 - E_0}{2} = \frac{\sqrt{4\Gamma^2 + J^2} + J}{2} = \frac{J}{2}[\sqrt{4\lambda^2 + 1} + 1], \qquad (2.83)$$

where $a = \sqrt{4\lambda^2 + 1} - 2\lambda$., defining the relevant variable $\lambda = \Gamma/J$.

The fixed points of the recurrence relation (rewritten in terms of λ) are

$$\lambda^* = 0 \qquad (2.84)$$
$$\lambda^* \to \infty \qquad (2.85)$$
$$\text{and} \quad \lambda^* \simeq 1.277. \qquad (2.86)$$

Now if the correlation length goes as

$$\xi \sim (\lambda - \lambda_c)^\nu, \qquad (2.87)$$

in the original system, then in the renormalized system we should have

$$\xi' \sim (\lambda' - \lambda_c)^\nu \qquad (2.88)$$

$$\Rightarrow \quad \frac{\xi'}{\xi} = \left(\frac{\lambda' - \lambda_c}{\lambda - \lambda_c}\right)^{-\nu} \Rightarrow \left(\frac{\xi'}{\xi}\right)^{-1/\nu} = \frac{d\lambda'}{d\lambda}\bigg|_{\lambda = \lambda_c \equiv \lambda^*}. \qquad (2.89)$$

Since the actual physical correlation length should remain same as we renormalize, ξ' (the correlation length in the renormalized length scale) must be smaller by the factor b (that scales the length), than ξ (correlation length in original scale). i.e., $\xi'/\xi = b$, or

$$b^{-1/\nu} = \left(\frac{d\lambda'}{d\lambda}\right)_{\lambda = \lambda_c = \lambda^*} \equiv \Omega \qquad \text{(say)}, \qquad (2.90)$$

hence, $\nu = \left(\dfrac{\ln \Omega}{\ln b}\right)_{\lambda=\lambda^*} = \dfrac{\ln \Omega}{\ln 2} \simeq 1.47,$ (for $b=2$), (2.91)

compared to the exact value $\nu = 1$. Similarly $E_g \sim \omega \sim (\text{time})^{-1} \sim \xi^{-z}$; $z = 1$. But for $b = 2$, we do not get $z = 1$. Instead, $\lambda'/\lambda \sim b^{-z}$ gives $z \simeq 0.55$. Energy gap

$$\Delta(\lambda) \sim |\lambda_c - \lambda|^s \sim \xi^{-z} \sim |\lambda^c - \lambda|^{\nu z} \tag{2.92}$$

$$\Rightarrow \quad s = \nu z.$$

Hence $s = 0.55 \times 1.47 \simeq 0.81$ (compared to the exact result $s = 1$). Results improve rapidly for large b values [6].

2.5 Equivalence of d-dimensional Quantum Systems and $(d+1)$-dimensional Classical Systems: Suzuki-Trotter Formalism

The Suzuki-Trotter formalism [7] is essentially a method for transforming a d-dimensional quantum Hamiltonian into a $(d+1)$-dimensional effective classical Hamiltonian giving the same canonical partition function. Let us illustrate this by applying it to the transverse Ising system. We start with the transverse Ising Hamiltonian

$$\mathcal{H} = -\Gamma \sum_{i=1}^{N} \sigma_i^x - \sum_{(i,j)} J_{ij} \sigma_i^z \sigma_j^z \tag{2.93}$$

$$= \mathcal{H}_0 + \mathcal{V} \tag{2.94}$$

The canonical partition function of $\mathcal{H}$ reads

$$Z = Tr e^{-\beta(\mathcal{H}_0 + \mathcal{V})}. \tag{2.95}$$

Now we apply the Trotter formula

$$\exp(A_1 + A_2) = \lim_{M \to \infty} [\exp A_1/M \exp A_2/M]^M, \tag{2.96}$$

even when $[A_1, A_2] \neq 0$. On application of this, Z reads

$$Z = \sum_i \lim_{M \to \infty} \langle s_i | [\exp(-\beta \mathcal{H}_0/M) \exp(-\beta \mathcal{V}/M)]^M | s_i \rangle. \tag{2.97}$$

Here s_i represent the i-th spin configuration of the whole system, and the above summation runs over all such possible configurations denoted by i. Now we introduce a number M of identity operators

$$\mathcal{I} = \sum_i^{2^N} |s_{i,k}\rangle\langle s_{i,k}|, \quad k = 1, 2, \ldots M. \tag{2.98}$$

in-between the product of M exponential in Z, and have

$$Z = \lim_{M\to\infty} Tr \prod_{k=1}^{M} \langle \sigma_{1,k} \ldots \sigma_{N,k}| \quad (2.99)$$
$$\times \exp\left(\frac{-\beta\mathcal{H}_0}{M}\right) \exp\left(\frac{-\beta\mathcal{V}}{M}\right) |\sigma_{1,k+1}\sigma_{2,k+1}..\sigma_{N,k+1}\rangle,$$

and periodic boundary conditions would imply $\sigma_{N+1,p} = \sigma_{1,p}$. Now,

$$\prod_{k=1}^{M} \langle \sigma_{1,k} \ldots \sigma_{N,k}| \exp\left(\frac{\beta}{M} \sum_{i,j} \sigma_i^z \sigma_j^z \right) |\sigma_{1,k+1} \ldots \sigma_{N,k+1}\rangle \quad (2.100)$$
$$= \exp\left[\sum_{i,j=1}^{N} \sum_{k=1}^{M} \frac{\beta J_{ij}}{M} \sigma_{i,k}\sigma_{j,k}\right],$$

where $\sigma_{i,k} = \pm 1$ are the eigenvalues of the σ^z operator. Also,

$$\prod_{k=1}^{M} \langle \sigma_{1,k} \ldots \sigma_{N,k}| \exp\left[\frac{\beta\Gamma}{M} \sum_i \sigma_i^x\right] |\sigma_{1,k+1} \ldots \sigma_{N,k+1}\rangle \quad (2.101)$$

$$= \left(\frac{1}{2}\sinh\left[\frac{2\beta\Gamma}{M}\right]\right)^{\frac{NM}{2}} \exp\left[\frac{1}{2}\ln\coth\left(\frac{\beta\Gamma}{M}\right) \sum_{i=1}^{N}\sum_{k=1}^{M} \sigma_{i,k}\sigma_{i,k+1}\right]. \quad (2.102)$$

The last step follows because
$$e^{a\sigma^x} = e^{-i(ia\sigma^x)} = \cos(ia\sigma^x) - i\sin(ia\sigma^x) = \cosh(a) + \sigma^x \sinh(a), \quad (2.103)$$
and then
$$\langle \sigma | e^{a\sigma^x} | \sigma' \rangle = \left[\frac{1}{2}\sinh(2a)\right]^{1/2} \exp\left[(\sigma\sigma'/2)\ln\coth(a)\right], \quad (2.104)$$
as
$$\langle \uparrow | e^{a\sigma^x} | \uparrow \rangle = \left[\frac{1}{2}\sinh(2a).\coth(a)\right]^{1/2} = \cosh(a) = \langle \downarrow | e^{a\sigma^x} | \downarrow \rangle \quad (2.105)$$
and
$$\langle \uparrow | e^{a\sigma^x} | \downarrow \rangle = \left[\frac{1}{2}\sinh(2a)/\coth(a)\right]^{1/2} = \sinh(a) = \langle \downarrow | e^{a\sigma^x} | \uparrow \rangle. \quad (2.106)$$

Thus the partition function reads
$$Z = C^{\frac{NM}{2}} Tr_\sigma(-\beta\mathcal{H}_{eff}[\sigma]) \quad ; \quad C = \frac{1}{2}\sinh\frac{2\beta\Gamma}{M} \quad (2.107)$$

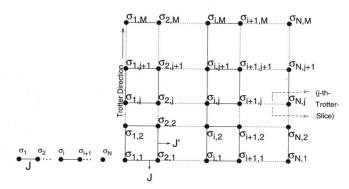

Figure 2.4: The Suzuki-Trotter equivalence of a quantum one dimensional chain and a (1+1) dimensional classical system. J' indicates the additional interaction in the Trotter direction.

where the effective classical Hamiltonian is

$$\mathcal{H}_{eff}(\sigma) = \sum_{(i,j)}^{N} \sum_{k=1}^{M} \left[-\frac{J_{ij}}{M} \sigma_{ik}\sigma_{jk} - \frac{\delta_{ij}}{2\beta} \ln \coth\left(\frac{\beta\Gamma}{M}\right) \sigma_{ik}\sigma_{ik+1} \right]. \quad (2.108)$$

The Hamiltonian $\mathcal{H}_{eff}$ is a classical one, since the variables $\sigma_{i,k}$'s involved are merely the eigenvalues of σ^z. Hence there is no non-commuting part in $\mathcal{H}_{eff}$. It may be noted from (2.108) that M should be at the order of $\hbar\beta$ (we have taken $\hbar = 1$ in the calculation) for a meaningful comparison of the interaction in the Trotter direction with that in the original Hamiltonian (see Fig. (2.4)). For $T \to 0$, $M \to \infty$, and the Hamiltonian represents a system of spins in a $(d+1)$-dimensional lattice, which is one dimension higher than the original d-dimensional Hamiltonian, as is evident from the appearance of one extra label k for each spin variable (see Fig. (2.4)). Thus corresponding to each single quantum spin variable σ_i in the original Hamiltonian we have an array of M number of classical replica spins σ_{ik}. This new (time-like) dimension along which these classical spins are spaced is known as the Trotter dimension. From the explicit form of $\mathcal{H}_{eff}$, we see that in addition to the previous interaction (J) term $(-\sum_{i,j}^{N} J_{ij}\sigma_i\sigma_j)$, there is an additional nearest neighbour interaction (J') between the Trotter replicas corresponding to the same original spin, along the Trotter direction, given by the term $(\sum_{i,j}^{N} \sum_{k=1}^{M} -\frac{\delta_{ij}}{2\beta} \ln \coth (\beta\Gamma/M)\sigma_{ik}\sigma_{iK+1})$ (as shown in Fig. 2.4). For finite temperature, the optimal width of the lattice in the Trotter direction is finite and the critical behaviour remains d-dimensional.

The calculations, and consequently the effective Hamiltonian in Eq. 2.108, are valid for any general interaction J_{ij}; of course, Γ has been taken to be nonrandom. Fig. (2.4) describes a situation where J_{ij} is nonrandom (we had $J_{ij} = J$). For random J_{ij}, we can see from Eq. 2.108 that the interactions in the Trotter direction remain identical (J') whereas the spatial randomness in interactions for various Trotter slices get correlated as indicated in Fig. (2.5).

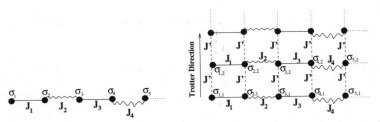

Figure 2.5: At the left is a portion of a one dimensional quantum Ising chain with random exchange interactions and at the right is a part of its Suzuki-Trotter equivalent classical lattice with randomness correlated in Trotter direction.

2.6 Transverse Ising Spin Glasses

2.6.1 Classical Spin Glasses: A Summary

Spin glasses are magnetic systems with randomly competing (frustrated) interactions [8]. Frustration is a situation where all of the spins present in the system cannot energetically satisfy every bond associated to them. Here the frustration arises due to competing (ferromagnetic and anti-ferromagnetic) quenched random interactions between the spins. As a result there arise huge barriers ($O(N)$, N = system size) in the free-energy landscape of the system. In the thermodynamic limit, the height of such barriers can go to infinity.

These barriers strongly separate different configurations of the system, so that once the system gets stuck in a deep valley in between two barriers, it practically gets trapped around that configuration for a macroscopically large time. Because of frustration, the ground state is largely degenerate with the degeneracy being of the order of $\exp(N)$. As discussed above, these different ground state configurations are often separated by $O(N)$ barriers, so that once the system settles down in one of them, it cannot visit the others equally often in the course of time, as predicted by the Boltzmann probability factor. The system thus becomes "non-ergodic" and may be described by a nontrivial order parameter distribution [8] in the thermodynamic limit (unlike the unfrustrated cooperative systems, where the distribution becomes trivially delta function-like). The spins in such a system thus get frozen in random orientations below a certain transition temperature. Although there is no long range magnetic order, i.e., the space average of spin moments vanishes, the time average of any spin is nonzero below the transition (spin-glass) temperature. This time average is treated as a measure of spin freezing or spin glass order parameter.

Several spin glass models have been studied extensively using both analytic and computer simulation techniques. The Hamiltonian for such models can be written as

$$\mathcal{H} = -\sum_{i<j} J_{ij} \sigma_i^z \sigma_j^z \qquad (2.109)$$

2.6. Transverse Ising Spin Glasses

where $S_i^z = \pm 1, 2, \ldots, N$, denote the Ising spins, interacting with random quenched interactions J_{ij}, which differs in various models. We will specifically consider three extensively studied models.

(a) In the Sherrington-Kirkpatrick (S-K) model J_{ij} are long-ranged and are distributed with a Gaussian probability (centered around zero), as given by

$$P(J_{ij}) = \left(\frac{N}{2\pi J^2}\right)^{1/2} \exp\left(\frac{-NJ_{ij}^2}{2J^2}\right) \qquad (2.110)$$

(b) In the Edwards-Anderson (EA) model, the J_{ij}'s are short-ranged (say, between the nearest neighbours only), but similarly distributed with Gaussian probability (2.110)

(c) In another model, the J_{ij}'s are again short-ranged, but having a binary ($\pm J$) distribution with probability p:

$$P(J_{ij}) = p\delta(J_{ij} - J) + (1-p)\delta(J_{ij} - J). \qquad (2.111)$$

The disorder in the spin system being quenched, one has to perform configurational averaging (denoted by overhead bar) over $\ln Z$, where $Z(= Tr \exp -\beta \mathcal{H})$ is the partition function of the system. To evaluate $\overline{\langle \ln Z \rangle}$, one usually employs the replica trick based on the representation $\ln Z = \lim_{n \to \infty}[(Z^n - 1)/n]$. Now, for a classical Hamiltonian (with all commuting spin components), $Z^n = \prod_{\alpha=1}^{n} Z_\alpha = Z(\sum_{\alpha=1}^{n} H_\alpha)$, where H_α is the α-th replica of the Hamiltonian $\mathcal{H}$ in Eq. 2.109 and Z_α is the corresponding partition function. The spin freezing can then be measured in terms of replica overlaps, and the Edwards-Anderson order parameter takes the form

$$q = \frac{1}{N}\sum_{i=1}^{N} \overline{\langle S_i^z(t) S_i^z(0)\rangle}|_{t \to \infty} \simeq \frac{1}{N}\sum_{i=1}^{N} \overline{\langle S_{i\alpha}^z S_{i\beta}^z \rangle}, \qquad (2.112)$$

where α and β corresponds to different replicas.

Extensive Monte Carlo studies, together with the analytical solutions for the mean field versions of the S-K and EA models, have revealed the nature of spin glass transition. It appears that the lower critical dimension d_l^c for EA model, below which transition ceases to occur (with transition temperature T_c becoming zero), is between 2 and 3: $2 < d_l^c < 3$. Thu upper critical dimension d_u^c, at and above which mean field results (e.g., those of S-K model) apply, appears to be 6: $d_u^c = 6$. Within these dimensions ($d_l^c < d < d_u^c$), spin glass transitions occur (for Hamiltonian's with short-ranged interactions) and the transition behaviour can be characterized by various exponents. Although the linear susceptibility shows a cusp at the transition point, the nonlinear susceptibility $\chi_{SG} = (1/N)\sum_r g(r)$, where $g(r) = (1/N)\sum_i \overline{(\langle S_i^z S_{i+r}^z\rangle)^2}$, diverges at the spin glass transition point:

$$\chi_{SG} \sim (T - T_c)^{-\gamma_c}, \quad g(r) \sim r^{-(d-2+\eta_c)} f\left(\frac{r}{\xi}\right); \quad \xi \sim |T - T_c|^{-\nu_c} \quad (2.113)$$

Here ξ denotes the correlation length which determines the length scaling in the spin correlation function $g(r)$ (f in $g(r)$ denotes the scaling function). Numerical simulation gives $\nu_c = 1.3 \pm 0.1$, 0.80 ± 0.15, $1/2$ and $\gamma_c = 2.9 \pm 0.5$, 1.8 ± 0.4, 1 for $d = 2, 3$ and 6 respectively for the values of exponents. One can define the characteristic relaxation time τ through the time dependence of spin autocorrelation

$$q(t) = \overline{\langle S_i^z(t) S_i^z(0) \rangle} \sim t^{-x} \tilde{q}\left(\frac{t}{\tau}\right); \quad \tau \sim \xi^z \sim |T - T_c|^{-\nu_c/z_c} \quad (2.114)$$

where $x = (d - 2 + \eta_c)/2z_c$, and z_c denotes the classical dynamical exponent. Numerical simulations give $z_c = 6.1 \pm 0.3$ and 4.8 ± 0.4 in $d = 3$ and 4 dimensions respectively. Of course, such large values of z_c (particularly in lower dimensions) also indicates the possibility of the failure of power law variation (2.114) of τ with $T - T_c$ and rather suggests a Vogel-Fulcher like variation: $\tau \sim \exp[A/(T - T_c)]$. In the $\pm J$ spin glasses (type (c) above), some exact results are known along the 'Nishimori Line' [8], and the nature of the phase transition there is precisely known.

2.6.2 Quantum Spin Glasses

Quantum spin glasses [9–14] have the interesting feature that the transition in randomly frustrated (competing) cooperatively interacting systems can be driven both by thermal fluctuations or by quantum fluctuations. Quantum spin glasses can be of two types: vector spin glasses introduced by Bray and Moore (see [4]), where of course quantum fluctuation cannot be tuned, or a classical spin glass perturbed by some tunable quantum fluctuations e.g., as induced by a non commutative transverse field [4, 14]. The amount of quantum fluctuation being tunable, this Transverse Ising Spin Glass (TISG) model is perhaps the simplest model in which the quantum effects in a random system can be and has been studied extensively and systematically [4, 14].

The interesting in such quantum spin glass models is about the possibility of tunneling through the (infinitely high) barriers of the free energy landscape in the classical spin glass models (e.g., S-K model) due to the quantum fluctuations induced by the transverse field. In classical systems, the crossing of an infinitely high barrier is impossible given thermal fluctuations at any finite temperature. But quantum fluctuations can make a system tunnel through such a barrier, if its width is infinitesimally small. Such barrier widths are actually seen to decrease with system size indicating to an ergodic (replica symmetric) picture for the free-energy landscape.

Sherrington-Kirkpatrick Model in a Transverse Field

The Sherrington-Kirkpatrick (S-K) model in presence of a non-commuting tunneling field, given by the Hamiltonian

$$\mathcal{H} = -\sum_{ij} J_{ij} \sigma_i^z \sigma_j^z - \Gamma \sum_i \sigma_i^x, \quad (2.115)$$

2.6. Transverse Ising Spin Glasses

where the distribution of coupling constants follows the Gaussian distribution

$$P(J_{ij}) = \left(\frac{N}{2\pi\Delta^2}\right)^{1/2} \exp\left(\frac{-NJ_{ij}^2}{a\Delta^2}\right) \qquad (2.116)$$

was first studied by Ishi and Yamamoto [9].

Phase Diagram

Several analytical studies have been made to obtain the phase diagram of the transverse Ising S-K model (giving in particular the zero-temperature critical field). The problem of S-K glass in transverse field becomes a nontrivial one due to the presence of non-commuting spin operators in the Hamiltonian. This leads to a dynamical frequency dependent (spin) self-interaction.

(i) Mean Field Estimates:

One can study an effective spin Hamiltonian for the above quantum many body system within the mean field framework. a systematic mean field theory for the above model was first carried out by Kopec (see e.g., [4]), using the thermofield dynamical approach and the short time approximation for the dynamical spin self-interaction. Before going into the discussion of this approach, we shall briefly review the replica-symmetric solution of the classical S-K model ($\Gamma = 0$) in a longitudinal field given by the Hamiltonian

$$\mathcal{H} = -\sum_{\langle ij \rangle} J_{ij}\sigma_i^z\sigma_j^z - h\sum \sigma_i^z \qquad (2.117)$$

where J_{ij} follows the Gaussian distribution given by (2.116).

Using the replica trick, one obtains for configuration averaged n-replicated partition function $\bar{Z}^n$, given by

$$\bar{Z}^n = \sum_{(\sigma_{i\alpha}=\pm 1)} \int_{-\infty}^{\infty} P(J_{ij})dJ_{ij} \exp\left[\beta\sum J_{ij}\sum \sigma_{i\alpha}^z\sigma_{j\alpha}^z + \beta h\sum \sigma_{i\alpha}^z\right]. \qquad (2.118)$$

Performing the Gaussian integral, using Hubbard-Stratonovich transformation and finally using the method of steepest descent to evaluate integrals for thermodynamically large system, one obtains the free energy per site f,

$$-\beta f = \lim_{n \to 0}\left[\frac{\beta\Delta^2}{4}\left(1 - \frac{1}{n}\sum_{\alpha,\beta} q_{\alpha,\beta}^2 + \frac{1}{n}\ln Tr(\exp L)\right)\right], \qquad (2.119)$$

where $L = (\beta J)^2 \sum_{\alpha,\beta} q_{\alpha\beta}\sigma_\alpha^z\sigma_\beta^z + \beta \sum_{\alpha=1}^n \sigma_\alpha^z$ and $q_{\alpha\beta}$ is self-consistently given by the saddle point condition $(\partial f/\partial q_{\alpha\beta}) = 0$. Considering the replica symmetric case ($q_{\alpha\beta} = q$), one finds

$$-\beta f = \frac{(\beta\Delta)^2}{2}(1-q) + \frac{1}{\sqrt{2\pi}}\int_{-\infty}^{\infty} dr\ e^{-\frac{r^2}{2}}\ln\left[2\cosh\{\beta h(r)\}\right] \qquad (2.120)$$

where r is the excess static noise arising from the random interaction J_{ij} and the spin glass order parameter q is self-consistently given by

$$q = \frac{1}{\sqrt{2\pi}} \int_{-\infty}^{\infty} dr \; e^{-\frac{r^2}{2}} \tanh^2\{\beta h(r)\} \qquad (2.121)$$

and $h(r) = \Delta\sqrt{qr} + h$ can be interpreted as a local molecular field acting on a site. Different sites have different fields because of disorder, and the effective distribution of $h(r)$ is Gaussian with mean 0 and variance $\Delta^2 q$.

At this point we can introduce quantum effect through a transverse field term $-\Gamma \sum_i \sigma_i^x$ (with longitudinal field $h = 0$). The effective single particle Hamiltonian in the transverse Ising quantum glass can be written as

$$\mathcal{H}_s = -h^z(r)\sigma^z - \Gamma\sigma^x, \qquad (2.122)$$

where $h_z(r)$, as mentioned earlier, is the effective field acting along the z direction arising due to nonzero value of the spin glass order parameter. Treating $h^z(r)$ and σ as classical vectors in pseudo-spin space, one can write the net effective field acting on each spin as

$$h_0(r) = h^z(r)\hat{z} - \Gamma\hat{x}; \quad |h_0(r)| = \sqrt{h^z(r)^2 + \Gamma^2}. \qquad (2.123)$$

One can now arrive at the mean field equation for the local magnetisation, given by

$$m(r) = p(r)\tanh[\beta h_0(r)]; \quad p(r) = \frac{|h^z(r)|}{|h_0(r)|}, \qquad (2.124)$$

and consequently, the spin glass order parameter can be written as

$$q = \frac{1}{\sqrt{2\pi}} \int_{-\infty}^{\infty} dr \; e^{-r^2/2} \tanh^2\{\beta h_0(r)\} p^2(r). \qquad (2.125)$$

The phase boundary can be found from the above expression by putting $q \to 0$ ($h^z(r) = J\sqrt{qr}$ and $h_0 = \Gamma$), when it gives

$$\frac{\Gamma}{\Delta} = \tanh\left(\frac{\Gamma}{k_B T}\right). \qquad (2.126)$$

From the above we get $\Gamma_c = J$. Ishi and Yamamoto used the "reaction field" technique to construct a "TAP" like equation for the free energy corresponding to the Hamiltonian (55) and perturbatively expanded the free energy in powers of Γ upto the order Γ^2 to obtain

$$k_B T_c = \Delta[1 - 0.23(\Gamma/\Delta)^2]. \qquad (2.127)$$

(ii) Monte Carlo Studies:

Several Monte Carlo studies have been performed [9, 14] for S-K spin glass in transverse field. Applying Suzuki-Trotter formulation (as discussed earlier) of effective partition function, one can obtain the effective classical Hamiltonian in Mth Trotter approximation as

$$\mathcal{H}_{eff} = -\frac{1}{M}\sum_{i,j=1}^{N}\sum_{k=1}^{M} J_{ij}\sigma_{ik}\sigma_{jk} - \frac{1}{2\beta}\ln\coth\left(\frac{\beta\Gamma}{M}\right)\sum_{i=1}^{N}\sum_{k=1}^{M}\sigma_{ik}\sigma_{ik+1} \quad (2.128)$$
$$-\frac{NM}{2}\ln\left[\frac{1}{2}\sinh\frac{2\beta\Gamma}{M}\right],$$

where σ_{ik} denotes the Ising spin defined on the lattice (i,k), i being the position in the original S-K model and k denoting the position in the additional Trotter dimension.

Ray et al. [10] took $\Gamma \ll J$ and their results indeed indicate a sharp lowering of $T_C(\Gamma)$. Such sharp fall of $T_c(\Gamma)$ with large Γ is obtained in almost all theoretical studies of the phase diagram of the model.

2.6.3 Edwards-Anderson Model in a Transverse Field

The Hamiltonian for the Edwards-Anderson spin glass in presence of transverse field is that given by (2.115), where the random interaction this time is restricted among the nearest neighbours and satisfies a Gaussian distribution with zero mean and variance J, as given by

$$P(J_{ij}) = \frac{1}{\sqrt{2\pi}}\exp\left(-\frac{J_{ij}^2}{2J^2}\right). \quad (2.129)$$

With $\Gamma = 0$, the above model represents the E-A model with order parameter $q = \overline{\langle\sigma_i^z\rangle^2} = 1$ (at $T = 0$). When the transverse field is introduced, q decreases, and at a critical value of the transverse field the order parameter vanishes. To study this quantum phase transition using quantum Monte Carlo techniques, one must remember that the ground state of a d-dimensional quantum model is equivalent to the free energy of a classical model with one added dimension which is the imaginary time (Trotter) dimension. The effective classical Hamiltonian can be written as

$$\mathcal{H} = \sum_{k}\sum_{ij} K_{ij}\sigma_{ik}\sigma_{jk} - \sum_{k}\sum_{i} K\sigma_{ik}\sigma_{ik+1}, \quad (2.130)$$

with

$$K_{ij} = \frac{\beta J_{ij}}{M}; \quad K = \frac{1}{2}\ln\coth\left(\frac{\beta\Gamma}{M}\right), \quad (2.131)$$

where σ_{ik} are classical Ising spins and (i,j) denotes the original d-dimensional lattice sites and $k = 1, 2, \ldots, M$ denotes a time slice. Although the equivalence between classical and the quantum model holds exactly in the limit $M \to \infty$, one can always make an optimum choice for M. The equivalent classical Hamiltonian has been studied using standard Monte Carlo technique. The numerical estimates of the phase diagram etc. are reviewed in detail in [14].

2.6.4 $\pm J$ Model in a Transverse Field

Here the spin glass is modeled with a transverse Ising system, having only nearest neighbour interactions, whose Hamiltonian is given by (2.115) and J_{ij} follows a binary distribution given by (2.111). Here J is taken positive and p is thus the concentration of anti-ferromagnetic $-J$ bonds in the system.

Results in One Dimensional Chain

Here first we show analytically that in a one dimensional transverse Ising Hamiltonian with uniform J and Γ, if we replace some J bonds by $-J$ bonds randomly, then the resulting Hamiltonian can be gauge transformed back to one with uniform J, and hence the critical field remains unchanged with randomness concentration. Similar result for one dimensional system with distributed J had been obtained earlier [11].

Let us take the one dimensional random bond transverse Ising Hamiltonian

$$\mathcal{H} = -\sum_i J_i \sigma_i^z \sigma_{i+1}^z - \sum \Gamma \sigma_i^x, \qquad (2.132)$$

where the transverse field Γ is uniform throughout the system, and J_i's are randomly chosen from the similar distribution as given in 2.111). Since the J_i's have same magnitude J all through and their randomness is only in their sign, we may write $J_i = J\mathrm{sgn}(J_i)$, and thus the Hamiltonian (2.132) takes the form

$$\mathcal{H} = -J \sum \mathrm{sgn}(J_i) \sigma_i^z \sigma_{i+1}^z - \Gamma \sum \sigma_i^x. \qquad (2.133)$$

Now let us define a new set of spin variables as below

$$\tilde{\sigma}_i^z = \sigma_i^z \prod_{k=1}^{i-1} \mathrm{sgn}(J_k) \qquad (2.134)$$

$$\tilde{\sigma}_i^x = \sigma_i^x \qquad (2.135)$$

$$\tilde{\sigma}_i^y = \sigma_i^y \prod_{k=1}^{i-1} \mathrm{sgn}(J_k). \qquad (2.136)$$

It is easy to see that $\tilde{\sigma}$'s satisfy the same commutation and anti-commutation relations as those of σ's and hence will exhibit exactly the same dynamical behaviour. Now,

$$\tilde{\sigma}_i^z \tilde{\sigma}_{i+1}^z = \sigma_i^z \sigma_{i+1}^z \left[\prod_{k=1}^{i-1} [\mathrm{sgn}(J_k)]^2 \right] \mathrm{sgn}(J_i), \qquad (2.137)$$

2.6. Transverse Ising Spin Glasses

or,

$$\tilde{\sigma}_i^z \tilde{\sigma}_{i+1}^z = \sigma_i^z \sigma_{i+1}^z \text{sgn}(J_i), \tag{2.138}$$

since $[\text{sgn}(J_k)]^2 = 1$. Thus in terms of the new spin variables, Hamiltonian (2.133) becomes

$$\mathcal{H} = -J \sum_i \tilde{\sigma}_i^z \tilde{\sigma}_{i+1}^z - \Gamma \sum_i \tilde{\sigma}_i^x. \tag{2.139}$$

The above Hamiltonian describes the same random system in terms of new variables, and yet, as one can see, it has in itself no randomness at all. One can use a Jordan-Wigner transformation in terms of $\tilde{\sigma}$'s and see that here also quantum phase transition occurs only at $\Gamma \geq \Gamma_c (= J)$, just as it occurs in a non-random Hamiltonian in σ's. A numerical study [12] on such one dimensional chain is also in agreement with the above result.

Results for Square Lattice

We consider now the same system (represented by Hamiltonian (2.139)) on a square lattice of size 3×3 with periodic boundary condition. We again calculate the ground state and the first excited state energy E_0 and E_1 respectively as functions of the transverse field Γ, for different values of p. Each value of E_0 and E_1 is averaged over at least 10 configurations for each $p \neq 0$. Apart from Δ, we also calculate $\chi = \partial^2 E_0 / \partial \Gamma^2$ and their variations with Γ as shown in Figs. (2.6) and (2.7) respectively.

These results here are severely constrained by the system size. The value of the pure ferromagnetic critical field $\Gamma_c(p = 0)$ is found here to be around 2.2, while the series results or cluster algorithms give the value to be around 3.0.

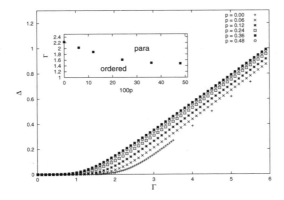

Figure 2.6: A numerical estimate of the configurational average of the energy gap Δ for a square lattice of size 3×3. The phase boundary obtained from $\Delta(\Gamma) = 0$ is outlined in the inset. (from [12])

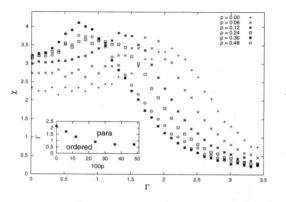

Figure 2.7: Here variation of $\chi = \partial^2 E_0/\partial \Gamma^2$ with Γ is shown. The transition point occurs at $\Gamma = \Gamma_c$ where χ diverges; for finite system one gets only a peak in χ at $\Gamma = \Gamma_c(p)$. We have outlined the corresponding phase boundary in the inset (from Ref. [12]).

This discrepancy is attributed to the smallness of our system size ($N = 3^2$). However the qualitative behaviour of the order-disorder phase boundary (between ferro/spin glass and para) seems to be reasonable: $\Gamma_c(p)$ decreases with p initially, and then increases again as p approaches unity (pure antiferromagnet). The use of periodic boundary condition here (to avoid some numerical errors) also restricts the domain features and thereby affects our results. The absence of the knowledge of the ground-state wave function (and the correlation functions) in this method prevents us from analysing the structure of the ordered phases.

2.6.5 Replica Symmetry in Quantum Spin Glasses

The question of existence of replica-symmetric ground states in quantum spin glasses has been studied extensively in recent years. Replica symmetry restoration is a quantum phenomena arising due to the quantum tunneling between the classically "trapped" states separated by infinitely high (but infinitesimally narrow) barriers in the free energy surface, which is possible as the tunneling probability is proportional to the barrier area, which remains finite.

To investigate this aspect of quantum glasses, one has to study the overlap distribution function $P(q)$ given by

$$P(q) = \overline{\sum_{l,l'} P_l P_{l'} \delta(q - q^{(ll')})}, \tag{2.140}$$

2.6. Transverse Ising Spin Glasses

where P_l is the Boltzmann weight associated with the state l and $q^{ll'}$ is the overlap between the sates l and l'

$$q^{(ll')} = \frac{1}{N}\sum_{i=1}^{N}\langle\sigma_i\rangle^{(l)}\langle\sigma_i\rangle^{(l')}. \tag{2.141}$$

One can also define the overlap distribution in the following form (for a finite system of size N)

$$P_N(q) = \overline{\langle \delta(q - q^{(12)})\rangle}, \tag{2.142}$$

where $q^{(12)}$ is the overlap between two sets of spins $\sigma_i^{(1)}$ and $\sigma_i^{(2)}$, with identical bond distribution but evolved with different dynamics,

$$q^{(12)} = \frac{1}{N}\sum_i \sigma_i^{(1)}\sigma_i^{(2)}. \tag{2.143}$$

$P_N(q) \to P(q)$ in the thermodynamic limit. In quantum glass problem one can study similarly this overlap distribution $P_N(q)$; and if the replica symmetric ground states exists, the above function must tend to a delta function in thermodynamic limit. In the paramagnetic phase, the distribution will approach a delta function at $q = 0$ for the infinite system.

Ray, Chakrabarti and Chakrabarti [10], performed Monte Carlo simulations, mapping the d-dimensional transverse S-K spin glass Hamiltonian to an equivalent $(d+1)$-dimensional classical Hamiltonian and addressed the question of stability of the replica symmetric solution, with the choice of order parameter distribution function given by

$$P_N(q) = \overline{\langle \delta\left(q - \frac{1}{NM}\sum_{i=1}^{N}\sum_{k=1}^{M}\sigma_{ik}^{(1)}\sigma_{ik}^{(2)}\right)\rangle}, \tag{2.144}$$

where, as mentioned earlier, subscripts (1) and (2) refer to the two identical samples but evolved through different Monte Carlo dynamics. It may be noted that a similar definition for q (involving overlaps in identical Trotter indices) was used by Guo et al. [11]. Lai and Goldschmidt performed Monte Carlo studies with larger system size ($N \leq 100$) and studied the order parameter distribution function

$$P_N(q) = \overline{\langle \delta\left(q - \frac{1}{N}\sum_{i=1}^{N}\sigma_{ik}^{(1)}\sigma_{ik'}^{(2)}\right)\rangle}, \tag{2.145}$$

where the overlap is taken between different (arbitrarily chosen) Trotter indices k and k'; $k \neq k'$. Their studies indicate that $P_N(q)$ does not depend upon the choice of k and k' (Trotter symmetry). Rieger and Young (see [4]) also defined $q^{(12)}$ in similar way ($q^{(12)} = (1/NM)\sum_N^i \sum_{kk'}^M)\sigma_{ik}^{(1)}\sigma_{ik'}^{(2)}$. There are

76 Chapter 2. Phase Transitions in Disordered Quantum Systems

striking differences between the results Lai and Goldschmidt obtained with the results of Ray et al [10]. For $\Gamma \ll \Gamma_c$, $P(q)$ is found to have (in [10]) an oscillatory dependence on q with a frequency linear in N (which is probably due to the formation of standing waves for identical Trotter overlaps). However, with increase in N, the amplitude of oscillation decreases and the magnitude of $P(q=0)$ decreases, indicating that $P(q)$ might go over to a delta function in thermodynamic limit. The envelope of this distribution function appears to have an increasing $P(q=0)$ value as the system size is increased. Ray et al [10] argued that the whole spin glass phase is replica symmetric due to quantum tunneling between the classical trap states. Lai and Goldschmidt on the other hand, do not find any oscillatory behaviour in $P(q)$. In contrary they get a replica symmetry breaking (RSB) in the whole spin glass phase from the nature of $P(q)$, which in this case, has a tail down to $q = 0$ even as N increases. According to them their results are different from Ray et al [10] because of different choices of the overlap function. Goldschmidt and Lai have also obtained replica symmetry breaking solution at first step RSB, and hence the phase diagram.

Büttner and Usadel (see e.g., Chakrabarti et al [4]), have shown that the replica symmetric solution is unstable for the effective classical Hamiltonian (2.108) and also estimated the order parameter and other thermodynamic quantities like susceptibility, internal energy and entropy by applying Parisi's replica symmetry breaking scheme to the above effective Hamiltonian. Using static approximation, Thirumalai et al (see [4]), found a stable replica symmetric solution in a small region close to the spin glass freezing temperature near the phase boundary. But as mentioned earlier, in the region close to the critical line, quantum fluctuations are subdued by the thermal fluctuations. Thus the restoration of replica symmetry, which is essentially a quantum effect, perhaps cannot be prominent there.

All these numerical studies are for the equivalent classical Hamiltonian, obtained by applying the Suzuki-Trotter formalism to the original quantum Hamiltonian, where the interactions are anisotropic in the spatial and Trotter direction and the interaction in the Trotter direction becomes singular in the limit $T \to 0$. Obviously one cannot extrapolate the finite temperature results to the zero temperature limit. The results of exact diagonalization of finite systems ($N \le 10$) at $T = 0$ itself do not indicate any qualitative difference in the behaviour of the (configuration average) mass gap Δ and the internal energy E_g from that of a ferromagnetic transverse Ising case, indicating the possibility that the system might become ergodic. On the other hand, the zero temperature distribution for the order parameter does not appear to go to a delta function with increasing N as is clearly found for the corresponding ferromagnet (random long range interaction without competition). In this case the order parameter distribution $P(q)$ is simply the number of ground state configurations having the order parameter value as q. This perhaps indicate broken ergodicity for small values of Γ. The order parameter distribution also shows oscillations similar to that obtained by Ray et al [10].

Kim and Kim [15] have very recently investigated the S-K model in transverse field using imaginary time replica formalism, under static approximation. They have shown that the replica-symmetric quantum spin glass phase is stable in most of the area of the spin glass phase, as have been argued by Ray et al, in contrary to the results of Lai et al and Thirumalai et al (see e.g., Chakrabarti et al [4]).

2.7 Quantum Annealing

2.7.1 Multivariable Optimization and Simulated Annealing

Multivarable optimization problems consist of finding the maximum or minimum values of a function (known as a cost function) of very many independent variables. A given set of values for the whole set of independent variables defines a configuration. The value of the cost function depends on the configurations, and one has to find the optimum configuration that minimizes or maximizes the cost function. The explicit evaluation of the cost function for all possible configurations in this context generally turns out to be absolutely impracticable for most systems. One can start from an arbitrary state and go on changing the configuration following some stochastic rule, unless an extremum is reached. For example, in a minimization problem, one may start from an arbitrary configuration, change the configuration according to some stochastic rule, evaluate the cost function of the changed configuration, and then compare its value with that of the original configuration. If the new cost function is lower, the change is retained, i.e. the new configuration is accepted. Otherwise the change is not accepted. Such steps may be repeated for times unless a minimum is reached. But in most cases of multivariable optimisation problem, there are many local extrema in the cost function landscape, and one cannot be sure that the extremum that has been reached is the global one.

Kirkpatrick et al [16] proposed a very ingenious physical solution to this mathematical problem, now known by the name simulated annealing. The basic underlying principle of simulated annealing is as follows. It is known that an ergodic physical system, at any finite temperature resides in the global minimum of its free energy. The minimum of the free energy is a thermodynamic macrostate corresponding to a maximum number of accessible microscopic configuration. Hence at thermal equilibrium an ergodic system explores its configuration space randomly with equal apriori probability of visiting any configuration, and consequently is found most of the time at one or other of the configurations that corresponds to the free energy minimum (since the number of configurations corresponding to such minimum is overwhelmingly large compared to that of any other macro state). Now if the system starts from an arbitrary macro-state (not the minimum of free energy) then due to thermal fluctuations it reaches the free energy minimum within some time τ known as the thermal relaxation time of the system. For an ergodic system (away from the critical point) this relaxation time increases linearly with system size (which is exponentially smaller a number compared to the corresponding number of all possible configurations). Hence if one follows the random dynamics of the thermal relaxation of a system, then one will be able to reach the minimum of cost function (zero temperature free energy) in a substantially smaller time. What one needs to do is to view the cost function E as the internal energy of some system and start from an arbitrary configuration. Then one changes the configuration according to some stochastic rule, just as before.

Now if the energy is lowered by the change, the change is accepted, but if it is not, the change is not thrown away with certainty. Instead it is accepted with a probability equal to the Boltzmann factor $e^{-\Delta E/k_B T}$, where $\Delta E = E_{\text{(after change)}} - E_{\text{(before change)}}$ (since this is the way how systems relax thermally to their free energy minimum). Temperature T here is an artificially introduced parameter which has a high value initially, and is reduced slowly as time goes on, finally tending towards zero. At zero temperature the free energy is nothing but the internal energy of the system, and thus at the end of the final stage of annealing the system can be expected to be found, with a very high probability, in a configuration that minimizes the internal energy (cost function).

However this simulated annealing technique can suffer severe set backs when the system is "non-ergodic", like the spin glasses we discussed earlier. In such cases configurations corresponding to minimum of the cost function are separated by $O(N)$ sized barriers, and at any finite temperature thermal fluctuations will take practically infinite time to relax the system to the global minimum crossing these barriers in thermodynamic limit $N \to \infty$.

2.7.2 Quantum Annealing

In recent times, quantum annealing has emerged as among the most efficient methods for solving optimisation problems, particularly for hard problems where the other known algorithms take an exponentially long time in the size of the system [17–20]. The basic idea of quantum annealing is to implement the effects of adding quantum fluctuations to the system to overcome large barriers. In order to incorporate such quantum fluctuations one adds a non-commutative quantum kinetic term (say $\mathcal{H}'(t)$) to the classical one (say $\mathcal{H}$) that has to be optimised. The entire system can now be expressed by the Hamiltonian,

$$\mathcal{H}_{tot} = \mathcal{H} + \mathcal{H}'(t) \qquad (2.146)$$

The kinetic term $\mathcal{H}'(t)$ helps the system to penetrate the barriers thus retrieving the system from local traps. Initially the tunneling term is kept much higher than the classical part and thus the ground state is trivial. The kinetic term is reduced adiabatically assuring that the system eventually settles into one of the eigen states of $\mathcal{H}$ (at $t \to \infty, \mathcal{H}'(t) \to 0$), which is hopefully the ground state. The role of the tunneling term is to make the apparently large barriers

2.7. Quantum Annealing

transparent to the system, allowing it to attain any configuration with a finite probability.

The key element of this method is the choice of the kinetic term which can be expressed as $\Gamma(t)\mathcal{H}_{kin}$, where Γ is the tunable parameter that controls the quantum fluctuations. Initially the value of Γ is kept high so that the kinetic term dominates. Following a certain annealing schedule this term is brought to zero and, as mentioned before, if the reduction is substantially slow, the system will always be at the instantaneous ground state of the evolving Hamiltonian $\mathcal{H}_{tot}$ (adiabatic theorem of quantum mechanics [21]). As Γ is finally brought to zero, $\mathcal{H}_{tot}$ will coincide with $\mathcal{H}$ and the desired ground state is reached.

The quantum annealing technique can be implemented in a number of ways:

1. Quantum Monte Carlo annealing
2. Quantum annealing using real time adiabatic evolution
3. Annealing of a kinetically constrained system
4. Experimental methods

In this chapter we will concentrate our discussion on the Quantum Monte Carlo technique, which can again be classified into three types: (a) Path integral Monte Carlo (b) Green's function Monte Carlo and the (b) zero temperature Monte Carlo. The basic idea behind the path-integral Monte Carlo (PIMC) lies in the Suzuki-Trotter formalism [7] which maps a quantum problem to a classical one. It transforms a d-dimensional quantum Hamiltonian into a $d+1$-dimensional effective classical Hamiltonian giving the same canonical partition function (discussed in detail earlier). Quantum annealing of a Hamiltonian $\mathcal{H}_{tot}$ using the PIMC method consists of mapping $\mathcal{H}_{tot}$ to its equivalent classical one and simulating it at some fixed low temperature, so that thermal fluctuations are small. The quantity that is measured is the residual energy $\epsilon_{res}(\tau) = \epsilon(\tau) - \epsilon_0$, with ϵ_0 the true ground-state energy of the finite system and $\epsilon(\tau)$ the final energy of the system after reducing the transverse field strength Γ from some large initial value to zero in a time τ. This method was first implemented by Kadowaki and Nishimori [22] in solving the TSP problem, and an extensive use of the technique was made by Santoro and Tosatti [23]. The application of PIMC to classical spin glass has been discussed in the next section.

The Green's function Monte Carlo algorithm effectively simulates the real-time evolution of the wave-function during annealing which often requires *a priori* knowledge of the wave-function [23]. But as such *a priori* knowledge is not available in a number of systems (random optimization problems), and the application of this algorithm seems to be very restricted.

The zero temperature transfer-matrix Monte Carlo samples the ground state of the instantaneous Hamiltonian (specified by the given value of the parameters at that instant) using a projective method, where the Hamiltonian matrix itself (a suitable linear transformation of the Hamiltonian that converts into a positive matrix, in practice) is viewed as the transfer matrix of a finite-temperature classical system of one higher dimension [24, 25]. The sparsity of the Hamiltonian matrix for systems with local kinetic energy terms however leaves the classical system highly constrained and thus difficult to simulate for large system sizes.

Quantum annealing by quantum adiabatic evolution was applied to random instances of an NP-Complete problem by Farhi et al [26]. The application of quantum annealing in a kinetically constrained system was studied by Das et al [27]. In an experimental realization of quantum annealing for glass-like dipolar magnets, Brooke et al [28] demonstrated its advantage over classical (simulated) annealing.

2.7.3 Ergodicity of Quantum Spin Glasses and Quantum Annealing

The nonergodicity problem makes the search of the ground state of a classical spin glass a computationally hard problem (no algorithm bounded by some polynomial in system size exists for such NP-hard problems). The problems of Simulated Annealing in spin-glass-like systems can be overridden (at least partially) by employing the method of quantum annealing [22, 28]. The basic idea is as follows. Firstly the problem has to be mapped to a corresponding physical problem, where the cost function is represented by some classical Hamiltonian of the form of (2.109). Then a non-commuting quantum tunneling term is introduced. The introduction of such a quantum tunneling is supposed to make the infinitely high (but infinitesimally thin) barriers transparent to the system, and it can now make transitions to different configurations trapped between such barriers, in the course of the annealing. In other words, it is expected that application of a quantum tunneling term will make the free energy landscape ergodic, and the system will consequently be able to visit any configuration with finite probability. Finally the quantum tunneling term is tuned to zero ($\Gamma \to 0$) to get back the classical Hamiltonian. It may be noted that the success of quantum annealing is directly connected to the replica symmetry restoration in quantum spin glass [10, 15] due to tunneling through barriers (see Fig. (2.8) and the discussion in the preceding section).

Here, the d-dimensional quantum Hamiltonian (55) (to be annealed) is mapped to the $(d+1)$-dimensional effective Hamiltonian

$$\mathcal{H}_{d+1} = -\sum_{k+1}^{M} \left(\sum_{i,j}^{N} J_{ij} \sigma_i^k \sigma_j^k + J' \sum_{i=1}^{N} \sigma_i^k \sigma_i^{k+1} \right), \qquad (2.147)$$

where

$$J' = -\frac{MT}{2} \ln \coth\left(\frac{\Gamma}{MT}\right) > 0 \qquad (2.148)$$

is the nearest neighbour ferromagnetic coupling in Trotter direction, between the Trotter replicas of the same spin. In the course of annealing, the temperature

2.8. Summary and Discussions

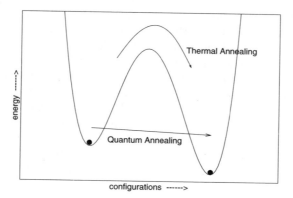

Figure 2.8: Schematic indication of the advantage of quantum annealing over classical annealing.

is kept constant at a low but nonzero value, and the tunneling field Γ is tuned slowly from a high initial value to zero. The decrease in Γ amounts to the increase in J' (as can be seen from above expression of J'). Initially at high Γ, J' is low, and each Trotter replica behaves almost like an independent classical spin system. The tunneling field is then lowered in small steps. In each such step, the system is annealed in presence of the small temperature. Finally as $\Gamma \to 0$, $J' \to \infty$, forcing all replicas to coincide at the end. As mentioned already, quantum annealing possibility directly rests on the replica symmetry restoration feature of quantum spin glasses, discussed in an earlier section.

2.8 Summary and Discussions

We have introduced the transverse Ising model in order to discuss the order-disorder transition (at zero temperature) driven by quantum fluctuations. Mean field theories are discussed next, where the application to BCS superconductivity theory is also discussed. Renormalization group techniques for the study of critical behaviour in such quantum systems are applied in the next section (for a chain). We have discussed the Suzuki-Trotter mapping of the d-dimensional quantum system to a $d+1$ dimensional classical system. We introduce the transverse Ising spin glass models, namely, the S-K model in a transverse field, the E-A model in a transverse field and the $\pm J$ model in a transverse field. The existing studies of their phase diagrams are discussed briefly. We then discuss the problem of replica symmetry restoration in quantum spin glasses. The application of the quantum annealing technique to access global minima in NP-hard problems is then discussed, as is the effectiveness of quantum tunneling over thermal barrier hopping.

It may be noted in this connection that some recent attempts have been made to apply similar annealing methods, induced by quantum fluctuations in transverse Ising models, for image restoration. Here relative advantages (or disadvantages) over conventional thermal annealing process may be compared in practical terms [29]. Like the near-global minima in free energy landscape of such spin glasses, the barriers are often globally contributed and barrier heights grow as the system size grows (unlike locally optimized configurations and the barriers between them). Classically, the system becomes non-ergodic due to these macroscopically high barriers (it is NP-hard to find the ground state), as thermal fluctuations are insufficient to permit the crossing of such barriers. Quantum tunneling does not necessarily require the scaling of barrier heights [10] to overcome them (as it operates by tunneling through; see Fig. (2.8)) thus helping to restore replica symmetry as well as annealing configurations [22, 28].

We are grateful to Amit Dutta, Anjan Kumar Chandra and Jun-ichi Inoue for many useful discussions and comments.

References

[1] R. Blinc, J. Phys. Chem. Solids **13** 204 (1960)

[2] P. G. de Gennes, Solid State Comm. **1** 132 (1963)

[3] R. Brout, K. A. Müller and H. Thomas, Solid State Comm. **4** 507 (1996); R. B. Stinchcombe, J. Phys. **C6** 2459 (1973)

[4] B. K. Chakrabarti, A. Dutta and P. Sen, *Quantum Ising Phases and Transitions in Transverse Ising Models*, Springer-Verlag, Heidelberg (1996); see also S. Sachdev, *Quantum Phase Transitions*, Cambridge Univ. Press, Cambridge (1999)

[5] P. W. Anderson, Phys. Rev. **112** 900 (1985); see also C. Kittel, *Quantum theory of Solids*, John Wiley, New York (1987)

[6] R. Jullien, P. Pfeuty, J. N. Fields and S. Doniach, Phys. Rev. B **18** 3568 (1978); R. Jullien, P. Pfeuty, J. N. Fields and K. A. Penson, in *Real Space Renormalization,* Ed. T. W. Brukhardt and J. M. J van Leeuween, Springer-Verlag, p.119 (1982)

[7] M. Suzuki, Prog. Theor. Phys. **46** 1337 (1971); *Quantum Monte Carlo Methods*, Ed. M. Suzuki, Springer-Verlag, Heidelberg (1987); R. J. Elloitt, P. Pfeuty and C. Wood, Phys. Rev. Lett. **25** 443 (1970)

[8] D. Chowdhury, *Spin Glass and Other Frustrated Systems*, World Scientific, Singapore (1986); H. Nishimori, *Statistical Physics of Spin Glasses and Information Processing: an Introduction,* Oxford University Press, Oxford (2001)

[9] B. K. Chakrabarti, Phys. Rev. B **24** 4062 (1981); H. Ishii and T. Yamamoto, J. Phys. C **18** 6225 (1985)

[10] P. Ray, B. K. Chakrabarti and A. Chakrabarti Phys. Rev. B **39** 11828 (1989)

[11] M. Guo, R. N. Bhatt and D. A. Huse, Phys. Rev. Lett. **72** 4137 (1994); D. Fisher, Phys. Rev. B **50** 3799 (1994)

[12] A. Das, A. Dutta and B. K. Chakrabarti, cond-mat/0310381

[13] W. Wu, B. Ellman, T. F. Rosenbaum, G. Aeppli and D. H. Reich, Phys. Rev. Lett. **67** 2076 (1991); W. Wu, D. Bitko, T. F. Rosenbaum and G. Aeppli, Phys. Rev. Lett. **74** 3041 (1993)

[14] R. N. Bhatt in *Spin Glasses and Random Fields*, Ed. A. P, Young, World Scientific, Singapore, pp. 225 - 249 (1998)

[15] D.- H. Kim and J.-J. Kim, Phys. Rev. B **66** 054432 (2002); see however, D. Thirumalai, Q. Li and T. R. Kirkpatrick, J. Phys. A **22** 3339 (1989)

[16] S. Kirkpatrick, C. D. Gelatt and M. P. Vecchi, Science, **220**, 671 (1983)

[17] A. Das and B. K. Chakrabarti, Rev. Mod. Phys. **80** 1061 (2008)

[18] A. Das and B. K. Chakrabarti (Eds.), *Quantum Annealing and Related Optimization Methods*, Lecture Note in Physics, Vol. **679** (Springer, Heidelberg, 2005)

[19] A. K. Chandra, A. Das and B. K. Chakrabarti (Eds.), *Quantum Quenching, Annealing and Computation*, Lecture Note in Physics, Vol. **802**, Springer, Heidelberg (2010)

[20] A. K. Chandra and B. K. Chakrabarti in *Computational Statistical Physics*, Ed. S. B. Santra and P. Ray (Hindustan Book Agency, New Delhi, 2011)

[21] M. S. Sarandy, L.-A. Wu and D. A. Lidar, *Quantum Inf. Process*, **3**, 331 (2004)

[22] T. Kadowaki and H. Nishimori, Phys. Rev. E **58** 5355 (1998)

[23] G. E. Santoro, R. Martonak, E. Tosatti and R. Car, *Science* **295**, 2427 (2002); R. Martonak, G. E. Santoro, and E. Tosatti, *Phys. Rev. B* **66**, 094203 (2002); R. Martonak, G. E. Santoro, and E. Tosatti, *Phys. Rev. E* **70**, 057701 (2004); D. Battaglia, G.E. Santoro, and E. Tosatti, *Phys. Rev. E* **71**, 066707 (2005); G. E. Santoro and E. Tosatti, *J. Phys. A* **39**, R393 (2006)

[24] J. P. Neirotti and M. J. de Oliviera, *Phys. Rev. B* **53**, 668 (1996)

[25] A. Das and B. K. Chakrabarti, *Phys. Rev. E* **78**, 061121 (2008)

[26] E. Farhi, J. Goldstone, S. Gutmann, J. Lapan, A. Lundgren and D. Preda, Science **292** 472 (2001)

[27] A. Das, B. K. Chakrabarti and R. Stinchcombe, *Phys. Rev. E* **72**, 026701 (2005)

[28] J. Brooke, D. Bitko, T. F. Rosenbaum and G. Aeppli, Science **284** 779 (1999); J. Brooke, T.F. Rosenbaum, and G. Aeppli, Nature **413** 610 (2001)

[29] H. Nishimori and K.Y.M. Wong, Phys. Rev. E **60** 132 (1999); J. Inoue, Phys. Rev. E **63** 046114 (2001); K. Tanaka, J. Phys. A **35** R81 (2002); J. Inoue, Physica Scr.T **106** 70 (2003)

Chapter 3

The Physics of Structural Glasses

Srikanth Sastry

A wide range of substances can exist as glasses, which are microscopically disordered, solid forms of matter. Typically, the glass state is reached by cooling a liquid to sufficiently low temperatures, and the analysis of the transformation to the glass is normally studied in this context. Glass formation occurs in the laboratory because the viscosity of the liquid becomes very high, causing the liquid to fall out of equilibrium on experimental time scales. Whether a true thermodynamic transition underlies the laboratory transformation is among the questions that remains to be answered. Another major theme is understanding the microscopic mechanisms leading to the dramatic growth of viscosity of liquids at low temperatures. A third theme is the understanding of relaxation processes that may occur in the glass state itself, or non-equilibrium relaxation that takes place en route to the glass state under conditions when the eventual state of a substance is the glass. Much of the study of the glass transition is focussed on the study of viscous liquids at low temperatures, as a key to understanding how the transformation to the glass state arises. An overview is given here of the phenomenology of the formation of glasses, computer simulation methods and results, and theoretical ideas which have been employed to probe the nature of relaxation in liquids.

3.1 Introduction

A wide variety of materials can exist at low temperatures in the form of a glass, which is *amorphous*, *i. e.*, a form of matter lacking in any long range structural order, but possessing mechanical properties of a solid [1–9]. Many familiar materials, in equilibrium, exist in a crystalline solid form at low temperatures. Solids are intuitively characterised by mechanical properties such as the retention of shape, and the fact that they do not flow. Crystalline solids further are characterised by a well defined, periodic microscopic arrangement of the atoms or molecules they are made of. Liquids are characterised by a lack of such microscopic order, densities comparable to crystalline solids, and the fact that they flow and are unable to retain their shape.

When a liquid is cooled, normally, it undergoes a first order phase transition at the freezing temperature to form a crystalline solid. However, a substance can remain in the liquid state even below the freezing temperature, in a *metastable* state. Such liquids are referred to as supercooled liquids. If a supercooled liquid is allowed to come to equilibrium, it does so, given sufficient time, to form a crystalline solid. However, if a supercooled liquid is cooled to lower temperatures sufficiently fast, it transforms to a distinct solid form, which is characterized the fact that its microscopic structure remains very close to that of the liquid from which it is formed. This *amorphous*, solid form of matter is termed a *glass*. Thus, for many materials, under conditions of observation (*e. g.* temperature), the glass state is a *metastable state*, and the state of the material in equilibrium is a crystalline solid. For some, however, notably polymers [10], the equilibrium crystalline structure, if it exists, is never attained and the glass is, for practical purposes, *the* state of the material at low temperatures.

Examples of glasses are abundant both in nature and among man-made materials: Most of the water in the universe is expected to be in the glassy form, in cometary tails [11]. A part of the earth's crust, in the form of amorphous silica formed during volcanic activity, is in the glassy state [12]. Window glass, whose dominant component is silica, is a man-made variant of the same material. A significant fraction of industrial plastics are glasses. Even among metals, which are almost exclusively crystalline solids, metallic glasses have been of interest of late for their unusual and desirable material properties [13].

In addition to the *route* to the glassy state described above, one may form amorphous solids in many other ways as well, *e. g.* depositing vapour on a cold substrate, or by application of pressure to a crystalline solid). The *conventional* method of preparing a glass, however, is to cool a liquid in a fashion that prevents crystallisation (in cases where the crystal forms the equilibrium state at low temperatures). Correspondingly, the term *glass transition* is used to designate the transformation of a liquid upon cooling to low temperatures into an amorphous solid. As described later, many interesting phenomena develop as a liquid is cooled towards the glass transition temperature, which bear directly on the glass state.

While understanding glasses as materials, with specific properties, require analysis of the synthesis, the chemical and physical properties of these amorphous solids at temperatures below the glass transition, general understanding of the nature of the glass state is also to be sought by studying materials in the liquid state, close to the glass transition. The body of scientific work outlined in this article falls in the category of studies which approach the glass transition from the 'liquid state'; the object of study is the glass-forming liquid at low temperatures, approaching the glass transition temperature from above.

Section II briefly describes some salient aspects of the phenomenology of the glass transition. Section III gives an overview of computer simulation methods that are typically employed, and the quantities calculated. The following section IV gives an overview of some of the theoretical approaches that have been used in analysing the approach to the glass state, with a focus on the Gibbs-DiMarzio-Adam entropy theory, mode coupling theory, approaches motivated by the analogy with the spin glass problem, and the energy landscape approach.

3.2 Phenomenology of Glass Formation

A brief overview of the phenomenology of glass formation are described in this section. Further details may be found in various recent reviews on the subject [1–9]. In particular, ref. [4] contains a comprehensive review of the experimental results and interpretations on most aspects of the structural glass transition problem.

A prerequisite for glass formation below the freezing temperature is that the substance does not crystallize. In supercooled liquids, crystallization takes place *via* the formation, through fluctuations, of crystallites that are large enough that they grow irreversibly [2, 14]. Two rates that are of importance are the rate of *nucleation* of crystalline nuclei that are larger than the critical size, and the growth rate of the nucleus once it forms. Both these rates exhibit a non-monotonic temperature dependence. The nucleation rate is determined by the free energy difference between the crystal and the metastable liquid (which favours the formation of the nuclei), the interface energy between the crystal nucleus and the mobility of the particles (which limit the rate).

If we consider the formation of a crystalline nucleus in the metastable liquid, the work of formation of such a nucleus is

$$\Delta G'(n) = c\,\sigma\,n^{2/3} + n\,\Delta G, \tag{3.1}$$

where c (and c' below) is a constant, σ is the interfacial energy per unit area, ΔG is the Gibbs free energy difference between the liquid and crystal per atom/molecule. The maximum value of this quantity, at a "critical nucleus size" $n^* \propto (\sigma/\Delta G)^3$ is

$$\Delta G'^* = \frac{c'\sigma^3}{\Delta G^2} \tag{3.2}$$

The nucleation rate is proportional to the probability of forming the critical nucleus, times a kinetic prefactor that determines the frequency of arrival of particles at the surface of the nucleus. The kinetic factor is proportional to the diffusivity of the particles.

Thus, we can write the nucleation rate as

$$I \propto D\,\exp\left[-\frac{\Delta G'^*}{k_B T}\right] \tag{3.3}$$

If we assume the excess enthalpy of the liquid to be roughly constant (as near the melting temperature) we can write the excess free energy as

$$\frac{\Delta G}{N} = \Delta h(1 - T/T_m) \tag{3.4}$$

and treating the diffusivity as varying in an Arrhenius fashion (in reality, the temperature dependence is stronger),

$$D(T) = D_0 \exp(-E/k_B T) \tag{3.5}$$

we get the nucleation rate, with $\Theta = T/T_m$, to be

$$I \propto \exp\left[-\frac{E'}{\Theta} + \frac{\Gamma}{\Theta(1-\Theta)^2}\right] \tag{3.6}$$

The nucleation rate is therefore nonmonotonic. Similarly, the growth rate of the nucleus also has a non-monotonic temperature dependence. There exists, thus, a temperature T_{max} below the freezing temperature for which the time taken for a given fraction of the system to crystallize is the least. This situation is illustrated in Fig. 3.1. If the liquid were to be cooled fast enough to a temperature below T_{max}, one has the possibility to study the supercooled liquid without interference of crystallization. The critical cooling rate is determined by the ratio of the difference of the freezing temperature and T_{max} and the transformation time for a small enough fraction that is at the limit of detection (say, one part per million). This critical cooling rate can be as high as $10^6 K/s$, e. g. for metallic glasses. The standard cooling rate that is adopted in most experimental studies is, however, $10K/min$. The *slowest* cooling rates that can be reached in computer simulations these days is about $10^8 K/s$ which are comparable to the fastest experimental cooling rates. For many glass formers, viscosity displays a strong temperature dependence, which deviates substantially from the Arrhenius temperature dependence, $\eta = \eta_0 \exp(E/k_B T)$ which

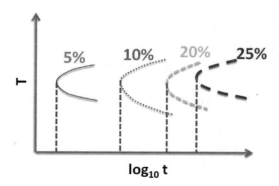

Figure 3.1: Schematic Time-Temperature-Transformation (TTT) diagram.

3.2. Phenomenology of Glass Formation

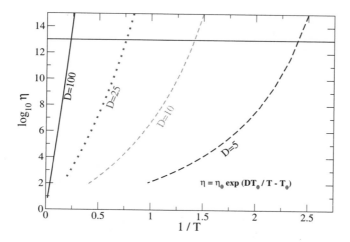

Figure 3.2: Variation of viscosity with temperature, shown schematically in an Arrhenius plot.

is used to describe the temperature dependence of many dynamical quantities in a variety of systems. The most familiar *glass*, silica, however exhibits close to Arrhenius temperature dependence. A sample of the variation of the viscosity with temperature is shown in Fig. 3.2, which also illustrates the range of materials which can form a glass. A generic functional form that appears more suitable is the Vogel-Fulcher-Tammann-Hesse (VFT) form, form,

$$\eta = \eta_0 \exp\left(\frac{A}{T - T_0}\right), \qquad (3.7)$$

although many instances are known where this is not a good description over the entire range of measurements. Further, various other functional forms have been proposed in the literature on varied grounds [4].

Such a form predicts the viscosity to become arbitrarily large, and diverge at a finite temperature T_0. However, when the viscosity reaches a value of around 10^{13} poise (which roughly corresponds to relaxation times of hundreds of seconds), liquids no longer achieve equilibrium under standard experimental conditions and time scales. This is a *kinetic* effect, and depends on the cooling rate, as seen later. Below the *transformation range* over which the liquid falls out of equilibrium, it transforms to a material with mechanical properties of a solid, namely, a glass. This transformation is what is referred to as the glass transition. Taking the glass transition temperature T_g as defined by the viscosity reaching a value of 10^{13} poise, we may inquire whether the plotting of the viscosity against a scaled temperature T/T_g yields a universal temperature dependence. This is not the case, as seen from data plotted in Fig. 3.3. Instead, a wide range of deviations from Arrhenius behaviour are observed. This deviation can be quantified in various ways, and is referred to as the *fragility* [15]

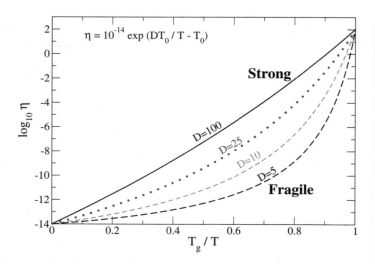

Figure 3.3: The "Angell plot" or fragility plot showing a schematic of the viscosity plotted against scaled temperature T/T_g. The liquids shown display a wide range of deviations from Arrhenius behaviour, which is quantified by the fragility of the glass former.

of the glass former. A large fragility implies a large deviation from Arrhenius behaviour, while a small fragility (displayed by liquids such as silica, which are referred to as 'strong' glass formers) implies a small deviation from Arrhenius behaviour. The falling out of equilibrium at the glass transition is also manifested in pronounced (but not discontinuous) changes in the temperature dependence of thermodynamic quantities such as the heat capacity and bulk density. The changes seen near the glass transition of the heat capacity and the volume are sketched schematically in Fig. 3.4 and 3.5. The heat capacity drops from high liquid state values to values very close to that of the crystal across the glass transition. The rate of change of the volume with temperature also shows a similar drop; equivalently, the coefficient of thermal expansion of the material drops to smaller values when the liquid undergoes a transformation to the glass. The kinetic nature of the glass transition can be probed, e. g., by considering the variation of volume with temperature at different cooling rates, both in the direction of decreasing and increasing temperatures. Fig. 3.6 shows schematically the variation of the volume for three different cooling rates. It is seen that the final volume reached at a temperature below the glass transition decreases with a decrease in the cooling rate, as does the glass transition temperature, which can be identified by where the steepness of the change of volume with temperature changes. Also shown are the volumes which would be observed if a glass obtained by cooling at rate γ_1 is heated up at a slower rate γ_2. In this case, one observes that the heating curve departs from the cooling curve, in the transformation range. In particular, the heating curve approaches

3.2. Phenomenology of Glass Formation

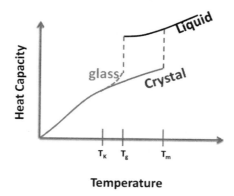

Figure 3.4: Variation of heat capacity with temperature, shown schematically, along with the heat capacity of the crystal. Note that the heat capacity drops in magnitude at the glass transition.

more closely what may be inferred to be the equilibrium behaviour by comparing the different curves. Similarly, heating a glass obtained through cooling at a slower rate γ_3, when heated at γ_2 shows deviations from the cooling curve in the opposite direction. From these, we may infer that the properties of the glass as well as its behaviour near the transformation range is subject to considerable history dependent effects. Since the heat capacity of a liquid is generally higher than that of the corresponding crystal, the entropy of the liquid decrease with temperature faster than that of the crystal, since the entropy is given by

$$S(T) = S(T_{ref}) + \int_{T_{ref}}^{T} \frac{C_p}{T'} dT' \qquad (3.8)$$

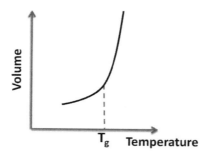

Figure 3.5: Variation of volume with temperature, shown schematically. Note that the change of volume with temperature shows a marked change across the glass transition.

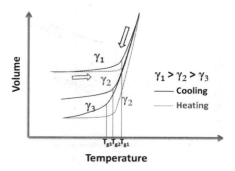

Figure 3.6: Variation of volume with temperature, shown schematically for cooling and heating at different rates. The full lines represent scans with decreasing temperature and the dashed lines represented scans with increasing temperature.

Consequently, the *excess* entropy of the liquid (relative to the crystal), by extrapolation, appears to vanish at a finite temperature, as first pointed out by Kauzmann [16], and referred to as the Kauzmann paradox. The observed behaviour of the excess entropy is shown in Fig. 3.7. While this situation is counter-intuitive, it does not disobey any laws of thermodynamics, as would, for example, a vanishing of the total entropy of the liquid at a finite temperature. In practice, the glass transition always occurs at a higher temperature

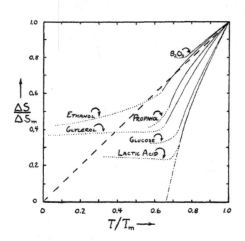

Figure 3.7: Variation of the excess entropy with temperature, suggesting a vanishing excess entropy at finite temperature. Reprinted with permission from Kauzmann W., *Chem. Rev.*, **43**, 219 (1948). Copyright (1948) American Chemical Society.

3.2. Phenomenology of Glass Formation

than the temperature of vanishing excess entropy. A theoretical resolution of this paradox, among other possibilities, involves the existence of an *ideal* glass transition, where the entropy associated with the degeneracy of *configurations* a liquid can assume vanishes, as described later. In this case, an understanding of the dramatic increase in properties such as the viscosity and other properties related to relaxation processes in the liquid must be sought in an understanding of the nature of this transition. However, the slow down of dynamics and the existence of a thermodynamic transition are in principle two independent issues, as emphasised by various authors (see, *e. g.*, [17]).

A second aspect of *slow dynamics* that becomes manifest as the glass transition temperature is approached is the time evolution of various relaxation processes. Relaxation phenomena are monitored using a number of experimental and theoretical tools, which generically involve the measurement of the time correlation of some quantity of interest, $C(t) = \langle \delta A(t) \delta A(0) \rangle$ where A is the observable of interest, and $\delta A = A - \langle A \rangle$ and the angular brackets $\langle \cdot \rangle$ indicate an ensemble average. A central correlation function of interest is the correlation of density fluctuations in the liquid (which is discussed more fully later). For the present, this function is shown schematically (as a function of log(t) as is conventionally done) in Fig. 3.8 to illustrate some key aspects. This correlation function decays rapidly at short times, in the regime referred to as the 'ballistic regime', followed by a plateau where the relaxation is very small. This intermediate time window is referred to, following the nomenclature in mode coupling theory, as the β regime. Oscillatory behaviour in the β regime are thought to be related to a feature called the 'Boson peak' seen, *e. g.* in Raman and neutron scattering, which we do not discuss further here (see [4] for details). The long time relaxation beyond the β regime is referred to as α relaxation, which at high temperatures is well described by exponential decay (Debye relaxation). However, at temperatures near the glass transition, this decay is slower than exponential, and is described by the Kohlrausch-Williams-Watts (KWW) or

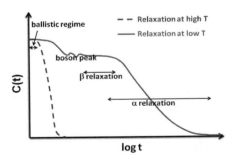

Figure 3.8: A schematic representation of the density-density time correlation function.

the 'stretched exponential' function:

$$C(t) = A\exp(-(t/\tau)^\beta). \quad (3.9)$$

Values of $\beta < 1$, which are obtained at low temperatures, imply that the relaxation cannot be viewed as being characterized by a single *exponential* process (or time scale), but is slower.

In liquids in normal conditions, the diffusion coefficient and the viscosity obey the well known Stokes-Einstein relation [18]:

$$D = \frac{k_B T}{q\pi a \eta} \quad (3.10)$$

where a is the molecular diameter and q is a constant equal to 3 or 2. At temperatures close to the glass transition, a breakdown of the Stokes-Einstein relation between the diffusion coefficient D and the viscosity η is observed [6, 19–22]. The measured diffusivities are considerably larger than the predictions based on measured viscosities. At the same time, rotational diffusion continues to be well correlated with viscosity, implying a decoupling of rotational and translational diffusion. These, and related observations are currently seen as manifestations of *heterogeneous dynamics* [6, 23, 24]– the distribution of relaxation times becomes sufficiently wide that quantities (such as viscosity and diffusion coefficient) which involve different averages over these relaxation times will begin to display qualitative differences with respect to temperature regimes where the relaxation times are narrowly distributed [25, 26, 109]. Ref. [6] provides a good overview of the results on this question. In particular, the picture is that the dynamics becomes *spatially* heterogeneous at low temperatures. Aspects of such spatially heterogeneous dynamics have also been studied by computer simulations, although at considerably higher temperatures than in experiments [28–30]. Among the observations of the simulation studies is the identification of a growing *dynamical* correlation length scale [31–35].

Another manifestation of the slow down of dynamics, which occurs both above and below the laboratory glass transition, is the phenomenon of *aging*– the glass forming liquid or glass continues to *anneal* on experimental times scales, and consequently, static as well as dynamic properties of the system continue to change over long time scales. The time translational invariance seen in systems in equilibrium is not observed. While this feature has been studied in the past [36], there has been considerable recent activity aimed at understanding such out of equilibrium behaviour. Detailed reviews of the recent literature may be found in [37], and more recently, the review by L. Cugliandolo in [38]. A brief introduction to aging has been given by G. Biroli [39]

3.3 Computer Simulations

Computer simulations have played a significant role in the study of liquids in general and in the elucidation of the behaviour of glass forming liquids in

3.3. Computer Simulations

particular. The reason for their utility is that, within the limitations described below, they provide complete microscopic information about the system being studied, which is very often not accessible by either experimental or other theoretical methods. The description of computer simulations here is meant to give an idea of how they work keeping in mind the case of a liquid made of particles interacting with a classical potential. The details as well as a more elaborate exposition can be found in a number of excellent textbooks, e. g. [40, 41].

The first issue that arises is the choice of the interaction potential. When one attempts to model a real material, e. g. water, this can be a source of error, since the actual interactions between molecules cannot be fully faithfully be represented by a classical potential. There has, however, been substantial amount of work in developing accurate enough potentials, and many of them do reasonably well. On the other hand, one may wish to study generic properties of the processes of interest, by employing a realistic enough potential, but which does not accurately represent any real material. One such popular choice is the Lennard-Jones potential,

$$V(r) = 4\epsilon \left[\left(\frac{\sigma}{r}\right)^{12} - \left(\frac{\sigma}{r}\right)^{6} \right] \qquad (3.11)$$

which models the interaction between noble elements (e. g. Argon). In the study of glass forming liquids, one typically employs mixtures of particles interacting with a potential such as the above, but with different ϵ and σ values for the different species, since the a single component Lennard-Jones liquid crystallizes too readily to be of use for studying a liquid exhibiting glassy behaviour at low temperatures.

Two standard methods for simulation are molecular dynamics and Monte Carlo. In both cases, the goal is to generate a series of atomic coordinates which forms a good approximate ensemble over which averages of desired properties may be calculated. In other words, for a property $A(\mathbf{r}, \mathbf{p})$ which depends on all atomic coordinates $\mathbf{r}$ and momenta $\mathbf{p}$, we wish to approximate its ensemble average:

$$\langle A \rangle \sim \frac{1}{N_s} \sum_i A(\mathbf{r}(i), \mathbf{p}(i)) \qquad (3.12)$$

where N_s is the number of entries in the series. The limitations to the goodness of the approximation arise from the fact that number of particles that can be simulated is necessarily finite (a few thousand particles, much smaller than Avogadro's number, which is the size of the real systems we wish to calculate the properties of), as is the number N_s (and once again, much smaller than the number of configurations over which a typical measurement may average the property in question). A part of the problem with the finite number of particles, namely the influence of the boundary on the properties of the system, are taken care of by imposing periodic boundary conditions, which may be visualized as a periodic replication of the volume being simulated indefinitely along each spatial dimension (this is not done explicitly in practice, of course, but methods of calculation are used to ensure the effect). Nevertheless, the finite size of the system can intrinsically alter the properties of the system, as seen, *e. g.* near a second order critical point, where correlation lengths crucial in determining the properties of the system grow boundlessly. In the simulation of glass forming liquids, the finite size effects do not appear to play such a crucial role, but they are there. However, they have not been characterized very well so far, since the length of the simulations needed pose a more immediate challenge to be tackled.

In the Monte Carlo method, the series of configurations generated are only of the coordinates of the particles and they are done in a manner that the series approximates the appropriate equilibrium ensemble. Therefore, there is no information regarding the dynamics, even though for Monte Carlo simulations with local moves (*i. e.* when adjacent configurations in the series are related to each other by a small displacement, typically a small change in the coordinates of one particle), one may think of the series generated as a time series and attempt to get some idea regarding the dynamics.

In molecular dynamics, instead, one explicitly calculates a time trajectory of the system of particles, by integrating the equations of motion for all the particles. Thus, one can calculate both equilibrium averages of time independent properties, as well as time dependent quantities. Given some initial coordinates and momenta, which are defined to be consistent with the desired properties of the system, one may use available algorithms to integrate the equations of motion in discrete steps. A commonly used algorithm is the velocity Verlet algorithm, given by:

$$\mathbf{r}_i(t + \delta t) = r_i(t) + \delta t \mathbf{v}_i(t) + \frac{\delta t^2}{2}\mathbf{F}_i(t) \qquad (3.13)$$

$$\mathbf{v}_i(t + \delta t) = v_i(t) + \frac{\delta t}{2}(\mathbf{F}_i(t) + \mathbf{F}_i(t + \delta t))$$

The use of this algorithm with generate a trajectory which will conserve the energy of the system and hence with produce an ensemble of configurations that are in the microcanonical ensemble. With modifications, one can also generate molecular dynamics trajectories that are at constant temperature (canonical) or constant temperature and pressure *etc* [40]. For the above algorithm to work reliably, the integration time step δt has to be small enough that a linear interpolation of the forces is accurate enough. One normally employs a time step of the order of a fs, and given that typical (long) simulations run to $\mathcal{O}(10^7)$ steps, run lengths of about 10 ns can be achieved. This is very far from $100s$ which is the time scale seen near the glass transition. However, such simulations allow the probing of $4-5$ decades of change in the relaxation times, which permits the analysis of many of the interesting questions concerning thermodynamic and dynamic properties of glass forming liquids.

3.4 Theoretical Approaches

In the following, a brief overview of some of the theoretical approaches that have been taken to explain the phenomena associated with the glass transition. These are the entropy theory due to Gibbs, Di Marzio and Adam, mode coupling theory, the energy landscape approach, and approaches motivated by analogy with spin glasses and theoretical ideas thereof. A number of other approaches, such as free volume theory, analysis of models of kinetic arrest in kinetically constrained models, frustration limited domain theory [42], *etc* are not covered, since a comprehensive review is not attempted here. Many of these themes are covered in various textbooks and recent reviews, *e. g.*, [2, 8, 9].

3.4.1 Gibbs, Di Marzio, Adam theory

The Kauzmann paradox suggests the possibility of a thermodynamic transition underlying the glass transition, which is related to the vanishing of the excess entropy of the liquid. One may imagine that the total entropy of the liquid arises from short time scale vibrational motion around some average structures (or configurations), and a *configurational* part, that arises from the fact that many such average arrangements of particles are possible in a liquid, unlike a crystal. For comparable densities and temperatures, one may expect the *vibrational* component of the entropy to be approximately the entropy of the crystal (which arises largely due to vibrations of particles around mean lattice positions). The excess entropy may then be seen as an approximation to the entropy of the liquid that arises from the possibility of having many distinct configurations. The vanishing of the excess entropy at the Kauzmann temperature may then be viewed as a consequence of the reduction and eventual vanishing of the number of distinct configurations available to the glass former. Gibbs and Di Marzio [43] developed a theory for the glass transition in polymer melts, using lattice statistics to calculate the partition function, in which the *ideal* glass transition (*i. e.* not the kinetically determined transformation seen in the laboratory) is a second order thermodynamic transition, associated with the vanishing of the entropy associated with the number of distinct polymer conformations.

A generalization of this work, extended to make contact with the nature of the dynamics approaching the glass transition, was attempted by Adam and Gibbs [44]. The connection to dynamics is made *via* the notion of 'cooperatively rearranging regions' (CRR). As the temperature of the liquid decreases (or, more easy to visualize, the density of the liquid increases) the atoms or molecules in liquid need to rearrange in a more and more cooperative fashion, in order to be mobile. Considering a CRR of size z, one may imagine calculating the partition function for it, and correspondingly the Gibbs free energy G. Noting that only a fraction of configurations of CRR allow rearrangement, we calculate a restricted partition function over these configurations, the corresponding free energy being G'. The free energy difference can be written as $z\delta\mu$ where $\delta\mu = (G' - G)/z$. The probability of rearrangement of a CRR is given by the Boltzmann weight associated with this free energy difference, or,

$$p(T) = A \, \exp(-\beta z \delta \mu). \tag{3.14}$$

For the entire system, the probability of rearrangement is dominated by the smallest available CRRs, and is given approximately by the same expression as above, with $z = z^*$ where z^* is the smallest CRR.

Now, if we consider the total configurational entropy of the system, S_c, it can be written in terms of the size of the CRR, as $S_c = (N/z)s_c$ where N is the total number of particles, (N/z is therefore the number of CRRs) and s_c is the configurational entropy per CRR which is assumed to be roughly constant. Since the relaxation time $\tau \propto p(T)^{-1}$, we can substitute for z in the above expression and write, for the relaxation times,

$$\tau(T) = A \, \exp\left[\frac{\beta \delta \mu s_c}{S_{conf}}\right] = A \exp\left[\frac{C}{TS_c}\right] \tag{3.15}$$

The above relation is referred to as the Adam-Gibbs relation, which has been found to be applicable in a number of experimental and simulation studies. Based on the discussion above, we may approximate $S_c \sim S_{liquid} - S_{crystal}$, i. e., the excess entropy ΔS. If we assume that the excess heat capacity of the liquid $\Delta C_P \sim K/T$, as verified in some cases, we obtain by integrating $\Delta C_P/T$, $T\Delta S = K(T/T_K - 1)$. Used in conjunction with the Adam-Gibbs relation above, would lead to the prediction

$$\tau = A \, \exp\left[\frac{C}{K(T/T_K - 1)}\right] \tag{3.16}$$

which is the VFT relation that describes experimental data reasonably in a number of cases. This correspondence would identify the Kauzmann temperature with the divergence temperature of the VFT equation.

While the Gibbs-Di Marzio-Adam theory has found wide application, there isn't a widely accepted and physical way to identify the CRR, although many suggestive results exist and the calculations themselves have been questioned [45] in the literature. Nevertheless, this theory forms the intuitive basis of much subsequent work, and continues to be used in the analysis of experimental results. We now turn to one of the approaches through which some of the ideas present in this theory have been sought to be concretized, namely the energy landscape approach.

3.4.2 The Energy Landscape Approach

The disordered structure of a liquid has the implication that the energies of interaction between particles will generally be very complicated, and that the part of configuration space explored by the material in the liquid state is characterised by the presence of many local minima of the potential energy. Such is the case also, e. g., in a crystalline solid if one allows for the presence of

3.4. Theoretical Approaches

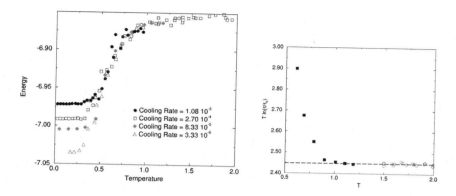

Figure 3.9: The top panel shows the average energy of local minima (inherent structures) sampled by a model liquid as a function of T, when the liquid is cooled at different rates. The temperature range over which the energies begin to deviate from near constant values to lower energies corresponds to a departure of the relaxation times from Arrhenius behavior, as shown in the lower panel. Reproduced from Sastry S., Debenedetti P. G. and Stillinger F. H., *Nature*, **393**, 554 (1998).

defects. The use of the phrase "energy landscape" to describe the complicated interactions in a glass forming liquid (and other disordered systems) therefore contains in addition the expectation that the complicated potential energy topography plays an essential role in determining the properties of the system. If such is the case, it is desirable to attempt a description of glass forming liquids in terms of quantities that define the nature of the potential energy landscape.

One way to get an idea of the relevance of the energy landscape to the dynamics of the liquid is to look for correlations between how the system samples the energy landscape and the nature of the dynamics, as a function of a control parameter such as the temperature. Fig. 3.9 shows results from molecular dynamics simulations [46] of a model liquid (the Kob-Andersen binary Lennard-Jones mixture [47]) at fixed density but decreasing the temperature at different rates. It is seen that the average energy of the local minima remains roughly constant at high temperatures, but begins to drop below a temperature of $T = 1$ (in reduced units [40]). Correspondingly, it is seen that the temperature dependent relaxation times deviate from an Arrhenius dependence ($\tau = \tau_0 \exp(E/k_B T)$; the rearrangement of terms shown in the figure should result in a constant E for Arrhenius behavior, and a T-dependent value otherwise). Intuitively, this makes sense - when the liquid begins to inhabit the basins of deeper and deeper energy minima, the process of relaxation involving the movement from one such basin to another of comparable depth, should become harder, leading to an increased effective activation energy. It is also noted that the dynamics becomes more complex, in that relaxation of density correlations becomes stretched exponential in character [46, 48], with the stretching exponent tracking the variation of the depth of the sampled minima [48]. Thus, below $T = 1$, terms the "onset temperature", one may expect the energy landscape picture to be appropriate in describing the properties of a supercooled liquid [49]. In the inherent structure approach [50], one considers the decomposition of the $3N$ dimensional (for an atomic liquid) configuration space of the liquid into basins of individual local potential energy minima, termed inherent structures (IS). A basin of a given minimum is defined as the set of points in the configuration space (or configurations) which map to that minimum under a local energy minimisation. The canonical partition function of the liquid can then be expressed as a sum over IS basins, the summand being partial partition functions defined for individual basins. In turn, the sum over basins is written in terms of (a) a distribution of minima in energy, and (b) the free energies of basins, as follows:

$$Q(N, \rho, T) = \Lambda^{-3N} \frac{1}{N} \int d\mathbf{r}^N exp\left[-\beta \Phi\right] \qquad (3.17)$$

$$= \sum_\alpha exp\left[-\beta \Phi_\alpha\right] \Lambda^{-3N} \int_{V_\alpha} d\mathbf{r}^N exp\left(-\beta(\Phi - \Phi_\alpha)\right)$$

$$\sim \int d\Phi \; \Omega(\Phi) \; exp\left[-\beta(\Phi + F_{vib}(\Phi, T))\right]$$

$$= \int d\Phi \; exp\left[-\beta(\Phi + F_{vib}(\Phi, T) - T\mathcal{S}_c(\Phi))\right]$$

where Φ is the total potential energy of the system, α indexes individual IS, Φ_α is the potential energy at the minimum, $\Omega(\Phi)$ is the number density of IS with energy Φ, and the configurational entropy density $\mathcal{S}_c \equiv k_B \ln \Omega$. The *basin free energy* $F_{vib}(\Phi_\alpha, T)$ is obtained by a restricted partition function sum over a given IS basin, V_α. Λ is the de Broglie wavelength, N is the number atoms in the liquid, T is the temperature, and ρ the density of the liquid. In the following, the dependence on ρ is not explicitly stated always since the interest is in T dependent behaviour at constant density. The configurational entropy of the liquid arises from the multiplicity of local potential energy minima sampled by the liquid at temperature T, and is related to the configurational entropy density above by

$$S_c(T) = \int d\Phi \; \mathcal{S}_c(\Phi) P(\Phi, T), \qquad (3.18)$$

where

$$P(\Phi, T) = \Omega(\Phi) \; exp\left[-\beta(\Phi + F_{vib}(\Phi, T))\right] / Q(N, \rho, T), \qquad (3.19)$$
$$= exp\left[-\beta(\Phi + F_{vib}(\Phi, T) - T\mathcal{S}_c(\Phi))\right] / Q(N, \rho, T),$$

is the probability density that IS of energy Φ are sampled at temperature T. In the above expression for the partition function, an assumption has been made that the basin free energy does not differ for different basins of the same IS energy. Without reference to the distribution of minima, the configurational

3.4. Theoretical Approaches

entropy can be defined as the difference of the total entropy of the liquid and the vibrational entropy of typical minima sampled at a given temperature:

$$S_c(\rho, T) = S_{total}(\rho, T) - S_{vib}(\rho, T). \qquad (3.20)$$

The "entropy theory" of Gibbs, Di Marzio and Adam [43, 44] define the ideal glass transition, underlying the laboratory transition, as an "entropy vanishing" transition where the configurational entropy vanishes (the configurational entropy is however not, as we have seen, defined in precisely the same way in [43, 44]). A similar picture also emerges from the study of mean field spin glass models and calculations motivated by them [51–56]. Whether such a transition exists for real materials is still a matter of debate [45, 57]. The calculations below produce such an entropy vanishing transition but it must be kept in mind that they result from extrapolations which may not be valid.

Further, Adam and Gibbs theory [44] relates the configurational entropy to relaxation times in the liquid:

$$\tau = \tau_0 \exp\left[\frac{A}{TS_c}\right], \qquad (3.21)$$

where A is a material specific constant. The validity of this relation has been verified by numerous experimental studies (which typically use the *excess* entropy of the liquid over the crystal in place of S_c) and computer simulation studies [58–61] (where configurational entropy is evaluated). Further, as remarked in the previous section, if S_c has the form $TS_c = K_{AG}(T/T_K - 1)$, the Adam-Gibbs relation results in the VFT form (3.7), which may be written as

$$\tau = \tau_0 \exp\left[\frac{1}{K_{VFT}(T/T_0 - 1)}\right], \qquad (3.22)$$

where T_0 is the temperature of apparent divergence of viscosity. K_{VFT} is a material specific parameter quantifying the *kinetic* fragility. Fragility is a measure of how rapidly the viscosity, relaxation times, *etc* of a liquid changes as the glassy state is approached [15]. Small values of K_{VFT} yield temperature dependence close to the Arrhenius form, while large values yield super-Arrhenius behaviour.

That the basin free energy F_{vib} arises from "vibrational" motion within individual basins is emphasised by the suffix *vib*. If this motion is sufficiently localised around the minima, a suitable procedure would be to approximate the basins as harmonic wells, and to evaluate the basin free energy within this approximation. The validity of such a procedure has been tested recently in various studies [49, 55, 60–63]. It is found that below the temperature where the liquid begins to exhibit aspects of *slow dynamics*, (non-Arrhenius behaviour of relaxation times, and stretched exponential relaxation) [46, 49, 62], a harmonic approximation of the basins is reasonable. However, this is not to be expected generally [59, 64], nor is it a requirement for calculating basin entropies. With a suitable criterion for defining inherent structure basins, one may also use constrained ensemble methods to evaluate the basin entropy [65, 66]. One must also consider whether results for classical systems typically studied in theoretical and computational studies hold for real systems.

In the harmonic approximation, the basin free energy is given simply in terms of the $3N$ vibrational frequencies that may be evaluated by diagonalizing the matrix of second derivatives, at a given local energy minimum, of the potential energy:

$$F_{vib} = k_B T \sum_{i=1}^{3N} \ln \frac{h\nu_i}{k_B T}, \tag{3.23}$$

or equivalently, the basin entropy,

$$\frac{S_{vib}}{k_B} = \sum_{i=1}^{3N} 1 - \log\left(\frac{h\nu_i}{k_B T}\right), \tag{3.24}$$

where ν_i are the vibrational frequencies of the given basin, and h is Plank's constant. From the form of S_{vib} it is apparent that the entropy difference between two basins arises solely due to the difference in their frequencies. Thus, such entropy differences remain finite as $T \to 0$ which is unphysical as the basin entropy of each basin and therefore their difference must go to zero for $T = 0$. An estimation of the effect of a quantum mechanical, as opposed to classical, treatment indicates that the effect of this artefact is not severe, if one considers deviations of the classical result at the glass transition temperature [67].

Calculations based on Eq.(3.24), where the vibrational frequencies are obtained numerically for energy minima generated in simulations, indicate [61] (see also [64, 68]) that the difference in S_{vib}, between basins is roughly linear in the basin energy. Thus one can write

$$\Delta S_{vib}(\Phi) \equiv S_{vib}(\Phi, T) - S_{vib}(\Phi_0, T) = \delta S \ (\Phi - \Phi_0), \tag{3.25}$$

and correspondingly,

$$F_{vib}(\Phi, T) = F_{vib}(\Phi_0, T) - T\delta S(\Phi - \Phi_0) \tag{3.26}$$

where Φ_0 is a reference basin energy. The latter expression follows since the internal energy $U_{vib} = 3Nk_B T$ for all basins.

In addition to the basin free energy, the partition function in Eq.(3.17) requires knowledge of the configurational entropy density $\mathcal{S}_c$. Various recent studies have explored methods for estimating $\mathcal{S}_c$ from computer simulations [61–63, 66, 69, 70]. It has been observed that the distribution $\Omega(\Phi)$ is well described by a Gaussian [61, 69, 71] (equivalently, $\mathcal{S}_c(\Phi)$ an inverted parabola). The arguments [69, 71] may not apply to low energy minima, nor are expected to be valid for all systems; indeed, a recent computational study of a model of silica [72] reveals a non-Gaussian $\Omega(\Phi)$, but this is related to a transition in silica from fragile to strong behaviour. Nevertheless, a Gaussian form for $\Omega(\Phi)$ allows for a straightforward evaluation of the partition function Eq.(3.17), and

3.4. Theoretical Approaches

whose validity has been tested in the range of temperatures where simulations are performed [49, 61].

The configurational entropy density is written as

$$\frac{\mathcal{S}_c(\Phi)}{Nk_B} = \alpha - \frac{(\Phi - \Phi_o)^2}{\sigma^2} \tag{3.27}$$

where α is the height of the parabola and determines the total number of configurational states, i.e. energy minima (the total number is proportional to $\exp(\alpha N)$), Φ_0 and σ^2 respectively define the mean and the variance of the distribution. The parameters α, Φ_0 and σ have been estimated from simulation data [61]. With the above form for $\mathcal{S}_c(\Phi)$ and Eq.(3.26) for the vibrational free energy, the partition function can be evaluated, from which the following temperature dependence of the configurational entropy, the ideal glass transition temperature T_K (defined by $\mathcal{S}_c(T_K) = 0$ and the IS energies are obtained:

$$\langle \Phi \rangle(T) = \Phi_0^{eff} - \frac{\sigma^2}{2Nk_BT}, \tag{3.28}$$

where $\Phi_0^{eff} = \Phi_0 + \frac{\sigma^2 \delta S}{2Nk_B}$,

$$TS_c(T) = K_{AG}^{PEL}(T) \ (T/T_K - 1); \ K_{AG}^{PEL}(T)$$
$$= \left(\frac{\sigma\sqrt{\alpha}}{2} + \frac{\sigma^2 \delta S}{4Nk_B}\right)\left(1 + \frac{T_K}{T}\right) - \frac{\sigma^2 \delta S}{2Nk_B}, \tag{3.29}$$

and

$$T_K = \sigma(2Nk_B\sqrt{\alpha} + \sigma\delta S)^{-1}. \tag{3.30}$$

These equations constitute relations that express quantities relevant to the thermodynamics of glass forming liquids, the configurational entropy and the ideal glass transition temperature, in terms of parameters that describe the "energy landscape" of the liquid, namely the distribution of local energy minima, and the topography of individual minima in the form of vibrational frequencies. In particular, the expressions for TS_c shows that the fragility of the liquid can be expressed in terms of parameters that quantify the "energy landscape" of the liquid. In particular, they show that when variations in basin entropy (with the energy of the minima) are unimportant, the spread in energy of the distribution of minima determines the fragility. When such variation is significant, it contributes to the fragility; if basin entropies (at the same temperature) are bigger for higher energy minima (which is the situation expected in the experimental, constant pressure conditions), there results a decrease in the fragility, as well as a decrease in the ideal glass transition temperature. The results here show that the use of excess entropy as a surrogate for the configurational entropy can be misleading. In view of the broad experimental test of the Adam-Gibbs relation, using the excess entropy, it has been suggested recently [73] that excess entropy and configurational entropy may vary proportionally. If one defines the excess entropy, not with respect to the crystal as is normally done, but with respect to the 'ideal glass' (*i. e.*, the lowest energy minimum), then one obtains an expression for the excess entropy as

$$TS_{ex}(T) = K_{ex}^{PEL}(T)(T/T_K - 1), \qquad (3.31)$$

where K_{ex} is the fragility index based on the excess entropy, given by,

$$K_{ex}^{PEL}(T) = \left(\frac{\sigma\sqrt{\alpha}}{2} + \frac{\sigma^2 \delta S}{4Nk_B}\right)\left(1 + \frac{T_K}{T}\right). \qquad (3.32)$$

Comparison of the expressions for K_{AG}^{PEL} and $K_{ex}^{PEL}(T)$ show that indeed they are closely related, but there is no proportionality in a strict sense. However, there are in principle subtle issues (*e. g.* in the analysis above T_K is used as a scaling temperature, but the existence of a Kauzmann temperature is itself an open question) that remain to be sorted out in how fragility is measured.

In the description above, the energy landscape was described in terms of a small number of parameters, α, σ, Φ_0, and δS, and we treated the system in the canonical ensemble, where the volume of the system was fixed and the temperature was varied. The values of these parameters should be expected to change as the volume or the density of the liquid changes. Incorporating such volume dependence to the parameters allows one to calculate the equation of state of the liquid in terms of the density dependence of these parameters, since the pressure is obtained as a volume derivative of the free energy. This approach has been used (see [76] for details and references) to analyse the "landscape" and thermal contributions to the equation of state, and to study out of equilibrium behaviour, and to analyse the existence of a second critical point in water [74].

In some liquids, notably silica, the temperature dependence of the diffusivity has been seen to show a crossover from super-Arrhenius to Arrhenius dependence. This crossover has been analysed in terms similar to the discussion above [72,75], and it has been shown that the energy landscape features of silica shows features that are consistent with expectations based on the changes of the dynamics, namely that the density of states of minima departs from a Gaussian at low energies, and correspondingly, the temperature dependence of the average inherent structure energy does not show a $1/T$ dependence.

As the above example of silica illustrates, the simplifying assumptions made in the discussion above leading to Eqs 22 - 24 should not be expected to hold for all systems. In addition to the shape of the density of states, the harmonicity of the basins may also not hold in some cases, but the corrections arising from such departures can be systematically included, at least in studies based on computer simulations. See [76] for some examples.

Many of the concepts and developments not dealt with in this introductory discussion may be found in the book by David Wales, "Energy Landscapes with applications to clusters, biomolecules and glasses" [77] which also contains, as the title indicates, applications to other systems than glass forming liquids. This includes the analysis of saddle points as a more elaborate description of the

3.4. Theoretical Approaches

energy landscape [78, 79], which is related to the instantaneous normal mode approach of Keyes and co-workers [80, 81]. A recent review by Cavagna [82] describes the analysis of saddle points in some detail.

An interesting recent energy landscape based approach to the study of dynamics in glass forming liquids is the analysis of *metabasins*. A metabasin is a collection of local energy minima, such that an escape from a metabasin constitutes an elementary step of structural relaxation, with the dynamics being uncorrelated between metabasins whereas the dynamics of transitions within a metabasin may be correlated [83, 84]. A recent review of the energy landscape approach by Heuer [85] describes in detail the analysis of metabasins.

3.4.3 Mode coupling theory

Mode coupling theory is a dynamical theory for the behavior of density correlation functions in a liquid, within the framework of the Mori-Zwanzig formalism that recasts the microscopic dynamics of an interacting many particle system in the form of a generalized Langevin equation. The quantity of interest here is the time correlation function of space and time dependent density,

$$\rho(\mathbf{r}, t) = \sum_i \delta(\mathbf{r} - \mathbf{r}_i(t)) \quad (3.33)$$

whose Fourier transform is,

$$\rho(\mathbf{q}, t) = \sum_i \exp[i\mathbf{q}.\mathbf{r}_i(t)]. \quad (3.34)$$

The density correlation function (also called the intermediate scattering function) is given by

$$F(q, t) = \frac{1}{N} \langle \rho(-\mathbf{q}, 0) \rho(\mathbf{q}, t) \rangle = \frac{1}{N} \sum_{i,j} \exp[i\mathbf{q}.(\mathbf{r}_i(t) - \mathbf{r}_j(0))]. \quad (3.35)$$

Similarly, the self part of the intermediate scattering function, also frequently studied, is defined to be

$$F_s(q, t) = \frac{1}{N} \sum_i \exp[i\mathbf{q}.(\mathbf{r}_i(t) - \mathbf{r}_i(0))]. \quad (3.36)$$

Keeping in mind these as the time correlation functions we wish to analyze, we first discuss the Mori-Zwanzig formalism for an unspecified set of functions, or operators, A. The discussion of the Mori-Zwanzig formalism and mode coupling theory follows closely the treatments in [86, 87].

Mori Zwanzig Formalism

Let A_ν be any operator that depends on the coordinates and momenta of a many body classical system specified by a Hamiltonian H. The time evolution of A_ν is given by

$$\frac{dA_\nu}{dt} = \{A_\nu, H\} \equiv iLA_\nu \tag{3.37}$$

(where L is the Liouville operator), whose symbolic solution is

$$A_\nu(t) = e^{iLt} A_\nu(0). \tag{3.38}$$

We consider in general a set of operators $A = \{A_1, ... A_N\}$.

We define a projection operator $\hat{P}$ into the subspace defined by these operators, as

$$\hat{P} B \equiv \sum_{A_i} \frac{|A_i\rangle \langle A_i| B\rangle}{\langle A| A\rangle} \tag{3.39}$$

The scalar product $\langle A_i| B\rangle$ is the ensemble average $\langle A_i^* B\rangle$. We have $\hat{P} A = A$. In defining the set of operators A one has in mind that these constitute a set of slow degrees of freedom, or modes, and the remaining degrees of freedom are fast, and act as a "bath" to these modes. It is instructive to keep in mind the simple example of a Brownian particle in a liquid where the degrees of freedom corresponding to the Brownian particle are the slow degrees of freedom, and the liquid degrees of freedom are the fast degrees of freedom. In this case, it is clear that the liquid degrees of freedom are the origin of the friction and the random forces to which a Brownian particle is subject.

Considering the time evolution of A, we can write,

$$\begin{aligned}
\frac{dA}{dt} &= e^{iLt}[\hat{P} + (1-\hat{P})]iLA(0) & (3.40)\\
&= e^{iLt}\hat{P} iLA(0) + e^{iLt}(1-\hat{P})iLA(0) & (3.41)\\
&= i\vec{\Omega}.\vec{A} + e^{iLt}(1-\hat{P})iLA(0) & (3.42)
\end{aligned}$$

Defining $G(t) \equiv e^{iLt}$ and $S(t)$ by

$$e^{iLt} = e^{iLt} S(t) + e^{i(1-\hat{P})Lt} \tag{3.43}$$

and taking the Laplace transform on both sides, we get

$$\begin{aligned}
G(s) &= \frac{1}{s-iL} & (3.44)\\
&= \frac{1}{s-i(\hat{P}+\hat{Q})L} \quad (\hat{Q} = 1-\hat{P}) & (3.45)\\
&= \frac{1}{s-i\hat{Q}L} + \frac{1}{s-iL} i\hat{P}L \frac{1}{s-i\hat{Q}L} & (3.46)
\end{aligned}$$

3.4. Theoretical Approaches

Taking inverse Laplace transform on both sides, we have

$$G(t) = e^{i\hat{Q}Lt} + \int_0^t d\tau e^{iL(t-\tau)} i\hat{P}L e^{i\hat{Q}L\tau}, \tag{3.47}$$

and hence

$$S(t) = e^{-iLt} \int_0^t d\tau e^{iL(t-\tau)} i\hat{P}L e^{i\hat{Q}L\tau} \tag{3.48}$$

$$= \int_0^t d\tau e^{-iL\tau} i\hat{P}L e^{i\hat{Q}L\tau}. \tag{3.49}$$

We have,

$$e^{iLt}(1-\hat{P})iLA = [e^{iLt} \int_0^t d\tau e^{-iL\tau} i\hat{P}L e^{i\hat{Q}L\tau} + e^{i(1-\hat{P})Lt}](1-\hat{P})iLA \tag{3.50}$$

$$= \int_0^t d\tau e^{iL(t-\tau)} i\hat{P}L f(\tau) + f(t) \tag{3.51}$$

where

$$f(t) = e^{i(1-\hat{P})Lt} i(1-\hat{P})LA(0) \tag{3.52}$$

$$= e^{(1-\hat{P})Lt} f(0) \tag{3.53}$$

Since $f(t)$ is always orthogonal to A that is $\overline{(A, f(t))} = 0$, we have

$$i\hat{P}Lf(\tau) = i(A, Lf(\tau))(A,A)^{-1}A \tag{3.54}$$

$$= (iLA, f(\tau)) \tag{3.55}$$

$$= i(i\hat{Q}LA, f(\tau)) \tag{3.56}$$

$$= -(f(0), f(\tau)). \tag{3.57}$$

We now define

$$K(t) \equiv (f(0), f(t))(A, A)^{-1} \tag{3.58}$$

With these relations, eqn. 3.42 becomes a generalized Langevin-like equation with a $f(t)$ taking the place of the "random" or fluctuation force.

$$\frac{dA}{dt} = i\vec{\Omega}.\vec{A} - \int_0^t d\tau K(\tau) A(t-\tau) + f(t) \tag{3.59}$$

If A is a time correlation function C then

$$\frac{dC}{dt} = i\vec{\Omega}.\vec{C} - \int_0^t d\tau K(\tau)C(t-\tau) + f(t) \tag{3.60}$$

Density Fluctuations

Let us consider A to be the operator for density fluctuations and longitudinal current in Fourier space.

$$A = \begin{pmatrix} \delta\rho_q \\ j_q^L \end{pmatrix}$$

And the correlation function are given by the matrix elements

$$C = \langle A^*(0)A(t) \rangle$$

Explicitly, we have,

$$\delta\rho_q = \sum e^{i\vec{q}.\vec{r}_i} - (2\pi)^3 \rho\delta(\vec{q}) \tag{3.61}$$

$$j_q^L = \frac{1}{m}\sum \vec{q}.\vec{p}_i e^{i\vec{q}.\vec{r}_i} \tag{3.62}$$

$$C(t) = \begin{pmatrix} \langle \delta\rho_{-q}(0)\delta\rho_q(t) \rangle & \langle \delta\rho_{-q}(0)j_q^L(t) \rangle \\ \langle j_{-q}^L(0)\delta\rho_q(t) \rangle & \langle j_{-q}^L(0)j_q^L(t) \rangle \end{pmatrix} \tag{3.63}$$

$$C(0) = \begin{pmatrix} NS(q) & 0 \\ 0 & \frac{Nk_BT}{m} \end{pmatrix} \tag{3.64}$$

Here, $S(q) = F(0)$, the structure factor. We note for use below that the lower left hand corner element of $C(t)$ above is $\frac{N}{iq}\frac{d^2F}{dt^2}$. We evaluate Ω by considering time $t=0$,

$$i\Omega = \langle A^*\dot{A}(0) \rangle \langle A, A \rangle^{-1} \tag{3.65}$$

$$= \begin{pmatrix} 0 & iq\frac{Nk_BT}{m} \\ iq\frac{Nk_BT}{m} & 0 \end{pmatrix} \begin{pmatrix} \frac{1}{NS(q)} & 0 \\ 0 & \frac{m}{Nk_BT} \end{pmatrix} \tag{3.66}$$

$$= \begin{pmatrix} 0 & iq \\ iq\frac{k_BT}{mS(q)} & 0 \end{pmatrix} \tag{3.67}$$

3.4. Theoretical Approaches

From this, the lower left hand corner element of $i\Omega C(0)$ is $\frac{iqNk_BT}{mS(q)}F$. Similarly, the fluctuating force at time $t=0$ is calculated as,

$$f(0) = (1-\hat{P})\dot{A} \tag{3.68}$$

$$= \begin{pmatrix} \delta\dot{\rho}_q \\ \frac{dj_q^L}{dt} \end{pmatrix} - \begin{pmatrix} 0 & iq \\ iq\frac{k_BT}{mS(q)} & 0 \end{pmatrix} \begin{pmatrix} \delta\rho_q \\ j_q^L \end{pmatrix} \tag{3.69}$$

$$= \begin{pmatrix} 0 \\ \frac{dj_q^L}{dt} - iq\frac{k_BT}{mS(q)}\delta\rho_q \end{pmatrix} \tag{3.70}$$

$$= \begin{pmatrix} 0 \\ R(q) \end{pmatrix}, \tag{3.71}$$

where the last step defines R_q. The correlation matrix of f is

$$K(t) = (f(0), f(t))\langle(A,A)^{-1}\rangle \tag{3.72}$$

$$= \begin{pmatrix} 0 & 0 \\ 0 & R_{-q}R_q(t) \end{pmatrix} \begin{pmatrix} \frac{1}{NS(q)} & 0 \\ 0 & \frac{m}{Nk_BT} \end{pmatrix} \tag{3.73}$$

$$= \begin{pmatrix} 0 & 0 \\ 0 & \frac{m\langle R_{-q}R_q(t)\rangle}{Nk_BT} \end{pmatrix} \tag{3.74}$$

Therefore, the product of K with C appearing in the memory term is

$$K(\tau)C(t-\tau) = \begin{pmatrix} 0 & 0 \\ 0 & \frac{m\langle R_{-q}R_q(t)\rangle}{Nk_BT} \end{pmatrix} \begin{pmatrix} \langle\delta\rho_{-q}(0)\delta\rho_q(t)\rangle & \langle\delta\rho_{-q}(0)j_q^L(t)\rangle \\ \langle j_{-q}^L(0)\delta\rho_q(t)\rangle & \langle j_{-q}^L(0)j_q^L(t)\rangle \end{pmatrix} \tag{3.75}$$

$$= \begin{pmatrix} 0 & 0 \\ \frac{m\langle R_{-q}R_q(t)\rangle}{Nk_BT}\langle j_{-q}^L(0)\delta\rho_q(t)\rangle & \frac{m\langle R_{-q}R_q(t)\rangle}{Nk_BT}\langle j_{-q}^L(0)j_q^L(t)\rangle \end{pmatrix} \tag{3.76}$$

Considering the lower left matrix element, we get the equation for $F(q,t)$, in the form

$$\frac{d^2F}{dt^2} + \frac{q^2k_BT}{mS(q)}F(q,t) + \frac{m}{Nk_BT}\int_0^t d\tau\langle R_{-q}R_q(t)\rangle\frac{dF}{dt} = 0 \tag{3.77}$$

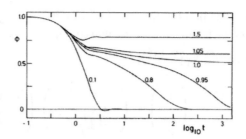

Figure 3.10: Solution for the density-density time correlation function using schematic mode coupling theory. Reprinted figure with permission from Leutheusser E., *Phys. Rev. A*, **29**, 2765 (1984). Copyright (1984) by the American Physical Society.

This equation is approximated in order to obtain the mode coupling equation. In the above equation, the quantity $\langle R_{-q}R_q(t)\rangle$ needs to be evaluated. An analysis of R_q reveals that it contains products of density operators in its expression, and therefore, contrary to expectation at the outset, the fluctuating force contains contributions from the slow modes, which leads to the coupling of modes in the final equation for F. Taking only the product of two densities as an approximation to R_q leads to a product of correlation functions F multiplying $\frac{dF}{dt}$ in the equation above. The approximation for $\langle R_{-q}R_q(t)\rangle$ used in mode coupling theory is

$$\langle R_{-q}R_q(t)\rangle = \frac{\rho^2(k_BT)^2}{2m^2}\sum_{\mathbf{k}}\left|\tilde{V}_{\mathbf{q-k,k}}\right|^2 F(k,t)F(|\mathbf{q-k}|,t) \qquad (3.78)$$

where

$$\tilde{V}_{\mathbf{q-k,k}} = \frac{ik_BT}{2mN}\left\{\frac{\hat{\mathbf{q}}\cdot\mathbf{k}}{S(k)} + \frac{\hat{\mathbf{q}}\cdot(\mathbf{q-k})}{S(|\mathbf{k-q}|)} - \mathbf{q}\cdot\hat{\mathbf{q}}\right\} \qquad (3.79)$$

Schematic mode coupling theory The mode coupling equation above has been studied in a schematic version, obtained by considering a single q-independent time correlation function, by various authors [88, 89]. We follow here the treatment in [89]. Considering a single time correlation function $\Phi(t)$ we write the schematic mode coupling equation for this function as

$$\frac{d^2\Phi}{dt^2} + \gamma\dot{\Phi}(t) + \Omega_0^2\Phi(t) + 4\lambda\Omega^2\int_0^t d\tau\,\Phi^2(\tau)\dot{\Phi}(t-\tau) = 0. \qquad (3.80)$$

Here, the parameter λ represents the strength of the mode coupling term, which one may expect to change as temperature in a supercooled liquid is varied. A Laplace transform of this equation can be written as

$$\Phi(z) = -\frac{1}{z - \frac{\Omega_0^2}{z+D(z)}} \qquad (3.81)$$

3.4. Theoretical Approaches

where
$$D(z) = i\gamma + 4\lambda\Omega^2 \mathcal{L}\{\Phi^2(t)\}. \tag{3.82}$$

Normally, the correlation function $\Phi(t)$ should decay to zero at long times. If the above equation were to describe a dynamical transition leading to structural arrest, however, one expects that the long time limit of $\Phi(t)$ will be non-zero. Writing an *ansatz* for $\Phi(t)$ based on this expectation as

$$\Phi(z) = -f/z + (1-f)\Phi_v(z), \tag{3.83}$$

where $\Phi_v(t)$ has an initial (t = 0) value of 1 and decays to zero at long time. $D(z)$ can be evaluated from this, and has the form

$$D(z) = -4\lambda^2 f^2/z + D_v(z) \tag{3.84}$$

Inserting this in the original expression for $\Phi(z)$ and comparing with the *ansatz*, one gets an expression for f as

$$f = \frac{1}{2}(1 + \sqrt{1 - 1/\lambda}). \tag{3.85}$$

For $\lambda > \lambda_c = 1$, this expression admits real solutions, and therefore, one has a non-decaying solution to $\Phi(t)$, as seen in Fig 3.10 which shows numerical solutions for Φ using the above equations.

The predicted dynamical transition is not seen in experiments. However, many of the observed predictions of mode coupling theory are found to be valid to varying degrees. The predicted mode coupling transition is a cross over but plays an important role in understanding the different regimes of dynamical slow down in glass forming liquids. Ref. [87] and the article by W. Kob in [90] describe in details the successes and failures of mode coupling theory. Another recent review which contains a detailed discussion of mode coupling theory is by S. P. Das [91]. A recent book by W. Gotze [92] is dedicated to a full exposition of mode coupling theory of dynamics in glass forming liquids.

Recently, an extension of mode coupling theory, *inhomogeneous mode coupling theory* (IMCT) [93] has been developed, which incorporates a description of spatially heterogeneous dynamics, mentioned earlier. IMCT has predictions relating to a growing length scale of spatially correlated dynamics, which has been studied by a number of authors [94–98]. Such a length scale also arises as a quantity of interest in the study of models studying "kinetic facilitation" as an approach to understanding glassy behavior [99, 100].

3.4.4 Spin glass theory

Considerable insight into the behavior of glass forming liquids has been gathered by the study of mean field spin glass models and the analysis of analogies with supercooled liquids [51–55]. An outline of important results is given here, and the reader is referred to various reviews [82, 101–103]. One of the models studied is the p-spin model, with the Hamiltonian

$$H = - \sum_{i_1 < i_2 ... < i_p} J_{i_1...i_p} \sigma_{i_1} ... \sigma_{i_p} - \sum_i h_i \sigma_i \qquad (3.86)$$

where p is the number of spins that interact via each term in the Hamiltonian, and $J_{i_1...i_p}$ are independent, Gaussian distributed random interactions. This model has been studied in particular in the version with soft spins (wherein the value of the spin is allowed to vary continuously with imposed addition terms in the Hamiltonian to keep the average length of the spins equal to one). For $p > 2$ this model exhibits a transition from a replica symmetric to one step replica symmetry broken solution at zero field. This transition has been identified with the entropy vanishing ideal glass transition at the Kauzmann temperature T_K. A study of the dynamics [51], however, reveals a dynamical transition at a higher temperature T_D, and the equation describing the decay of spin-spin correlations above such a temperature are identical in structure (other than the inertial - second time derivative - term) to the schematic mode coupling equation described above. Between T_D and T_K one has the presence of a multiplicity of metastable states, with a broad analogy to inherent structures (or metabasins), whose degeneracy vanishes at the Kauzmann temperature. Based on these results, the expected picture in the non-mean field situation is that the mode coupling description is operative at temperatures well above T_D, whereas at lower temperatures, the dynamics is dictated by "activated" transitions between the metastable states, which are expected to have finite barriers between them when the interactions are finite ranged. Support for such a picture have been obtained from computer simulations, following up a suggestion from M. Goldstein many years ago [104], analyzing the properties of inherent structures [105] and saddle points [78, 79].

In pursuing the analogy further, it has been proposed [53, 103] that there exists a length scale beyond which it becomes entropically favorable for regions of a liquid initially in one of the metastable states to make a transition to any one of the equivalent states. The free energy drive for such a transition is the configurational entropy associated with the most probable metastable states at the given temperature, whereas the hindrance arises from the "surface tension" between different structures that correspond to the different metastable states. In analogy with the nucleation problem discussed earlier, one may write

$$\Delta F(r) = c \, \sigma \, r^\theta + r^d \, (-TS_c) \qquad (3.87)$$

where r is the linear dimension of the rearranging region (we can write $r = n^{(1/d)}$ where n is the number of molecules in such a region). The free energy barrier, occuring for a characteristic size of the rearranging region, the so-called "mosaic length" ξ,

$$\xi \sim \left(\frac{1}{TS_c}\right)^{\frac{1}{d-\theta}} \qquad (3.88)$$

3.4. Theoretical Approaches

is

$$\Delta F \sim \left(\frac{1}{TS_c}\right)^{\frac{\theta}{d-\theta}} \qquad (3.89)$$

It has been argued [53, 103] that the surface tension exponent should depend on spatial dimensionality as $d/2$. If this is the case, using an activation expression for relaxation times, one recovers the Adam-Gibbs relation, which is also independent of spatial dimensionality. In addition to dynamical heterogeneity mentioned earlier, the mosaic length offers another length scale that is considered to be central to understanding the nature of slow relaxation. There has been considerable activity in understanding the role of relevant length scales in understanding relaxation behavior in glass forming liquids in recent years [106–112]. Some of the recent work on attempts to evaluate the mosaic length in simulations is reviewed in [82].

Shiladitya Sengupta is gratefully acknowledged for help in preparing the manuscript, and for proof reading the final version

References

[1] Review articles in *Science* **267**, (1995).

[2] Debenedetti, P. G., *Metastable Liquids* (Princeton University Press, Princeton, 1996).

[3] Ediger, M. D., Angell, C. A., and Nagel, S. R., *J. Phys. Chem.*, **100**, 13200-13212 (1996).

[4] Angell, C. A., Ngai, K. L., McKenna, G. B., McMillan, P. F. and Martin, S. W. *J. Appl. Phys.*, **88**, 3113-3157 (2000).

[5] Proceedings of *Unifying Concepts in Glass Physics, Trieste, 1999*, J. Phys. Cond. Mat., **12** (2000).

[6] Ediger, M. D., *Annu. Rev. Phys. Chem.*, **51**, 99–128 (2000).

[7] Debenedetti, P. G. and Stillinger, F. H., *Nature*, **410**, 259-267 (2001).

[8] Donth, E. *The Glass Transition* (Springer, Berlin, 2001).

[9] *Slow relaxations and nonequilibrium dynamics in condensed matter*, Barrat, J.-L, Feigelman, M., Kurchan J. and Dalibard, J. eds. (Springer, Berlin, 2003). See in particular the lectures by W. Kob (available also at http://www.arXiv.org/cond-mat/0212344), L. Cugliandolo (available also at http://www.arXiv.org/cond-mat/0210312) and G. Parisi (available also at http://www.arXiv.org/cond-mat/0301157).

[10] DiMarzio, E. A., *Ann. N.Y. Acad. Sci.*, **371**, 1-20 (1981).

[11] Jenniskens, P. and Blake, D. F. *Science*, **265**, 753-756 (1994).

[12] Bose, M. K., *Igneous Petrology* (World Press, Calcutta, 1997).

[13] Greer, A. L., *Science*, **267**, 1947-1953 (1995).

[14] Kelton, K. F. *Solid State Physics* (F. Seitz and D. Turnbull, eds.) **45** 75-177 (1991).

[15] Angell, C. A., *J. Non-Cryst. Solids*, **131-133**, 13-31 (1991).

[16] Kauzmann, W., *Chem. Rev.*, **43**, 219–256 (1948).

[17] Kurchan, J., *C. R. Acad. Sci. Paris, t.2, Serie IV*, 239-247 (2001); http://arXiv.org/abs/cond-mat/0011110

[18] Hansen, J.-P. and McDonald, I.R., Theory of Simple Liquids (3rd Ed.), Elsevier (2008).

[19] Fujara, F., Geil, B., Sillescu, H., Fleischer G., *Z. Phys. B.* **88** 195-204 (1992).

[20] Chang, I., Fujara, F., Geil, B., Heuberger, G. and Silescu, H., *J. Non-Cryst. Solids*, **172-174**, 248–255 (1994).

[21] Cicerone, M. T., Blackburn, F. R., and Ediger, M. D., *J. Chem. Phys.*, **102**, 471–479 (1995).

[22] Heuberger, G. and Silescu, H., *J. Phys. Chem.*, **100**, 15255–15260 (1996).

[23] Russel, E. V. and Israeloff, N. E., *Nature*, **408**, 695-698 (2000).

[24] Weeks, E. R., Crocker, J. L., Levitt, A. C., Schofielf, A. and Weitz, D. A., *Science*, **287**, 627-629 (2000).

[25] Tarjus, G. and Kivelson, D., *J. Chem. Phys.*, **103**, 3071-3073 (1995).

[26] Stillinger, F. H. and Hodgdon, J. A., *Phys. Rev. E.*, **50**, 2064-2068 (1994).

[27] Bhattacharya, S. and Bagchi, B., *J. Chem. Phys.*, **107**, 5852-5862 (1997).

[28] Hurley, M. M. and Harrowell, P. *Phys. Rev. E.*, **52**, 1694-1698 (1995).

[29] Yamamoto R., Onuki A., *J. Phys. Soc. Jpn.* **66** 25452548 (1997).

[30] Donati, C., Douglas, J. F., Kob, W., Plimpton, S. J., Poole, P. H., and Glotzer, S. C., *Phys. Rev. Lett.*, **80**, 2338-2341 (1998).

[31] Dasgupta, C., Indrani, A. V., Ramaswamy, S. and Phani, M. K., *Europhys. Lett.*, **15**, 307-312 (1991).

[32] Bennemann, C., Donati, C., Baschnagel, J., Glotzer, S. C., *Nature*, **399** 246-249 (1999).

[33] Glotzer, S. C., Novikov, V. N. and Schroder, T. B. Time-dependent, *J. Chem. Phys.* **112**, 509512 (2000).

[34] Franz, S. and Parisi, G., *J. Phys. Cond. Mat.*, **12**, 6335-6342 (2000).

[35] Donati, C., Franz, S., Glotzer, S. C. and Parisi, G., *J. Non-Cryst. Solids* **307**, 215224 (2002).

[36] Struick, L. C. *Physical aging in amorphous polymers and other materials* (Elsevier, Amsterdam 1978).

[37] Bouchaud, J.-P., Cugliandolo, L. F., Kurchan, J. and Mezard, M. in *Spin glasses and random fields* Young, A. P. ed. (World Scientific, Singapore, 1998)

[38] Cugliandolo, L., in *Slow relaxations and nonequilibrium dynamics in condensed matter*, J.-L. Barrat, M. Feigelman, J. Kurchan and J. Dalibard, eds. (Springer, Berlin, 2003) (available also at http://www.arXiv.org/cond-mat/0210312).

[39] Biroli, G., Lecture notes for the School on "Unifying Concepts in Glass Physics III" Bangalore, June 2004, *J. Stat. Mech.* P05014 (2005).

[40] Allen, M.P. and Tildesley, D.J., *Computer Simulations of Liquids*, (Clarendon Press, Oxford, 1987).

[41] Frenkel, D. and Smit, B., *Understanding Molecular Simulation: From Algorithms to Applications*, (Academic Press, San Diego, 1996).

[42] Kivelson, S. A., Kivelson, D., Zhao, X., Nussinov, Z. and Tarjus, G., *Physica A*, **219**, 27–38 (1995); Tarjus, G., Kivelson, D. and Viot, P., *J. Phys. Cond. Mat.*, **12**, 6497-6508 (2000).

[43] Gibbs, J. H. and Di Marzio, E. A. *J. Chem. Phys.*, **28**, 373-383 (1958).

[44] Adam, G. and Gibbs, J. H., *J. Chem. Phys.*, **43**, 139-146 (1965).

[45] Gujrati, P.D. *J. Phys. A*, **13**, L437 (1980) ; Gujrati, P.D. *J. Stat. Phys.* **28**, 441 (1982); P. D. Gujrati and M. Goldstein, it J. Chem. Phys. **74**, 2596 (1981).

[46] Sastry, S., Debenedetti, P. G. and Stillinger, F. H., *Nature*, **393**, 554-557 (1998).

[47] Kob, W. and Andersen, H. C., *Phys. Rev. E*, **51**, 4626-4641 (1995); Vollmayr, K., Kob, W. and Binder, K. *J. Chem. Phys.*, **105**, 4714-4728 (1996).

[48] Sastry, S., Debenedetti, P. G., Stillinger, F. H.,Schröder, T. B., Dyre, J. C., and Glotzer, S. C., *Physica A*, 270, 301-308 (1999).

[49] Sastry, S., *PhysChemComm*, 14, (2000);(http://arXiv.org/abs/cond-mat/0012054).

[50] Stillinger, F. H. and Weber, T. A., *Phys. Rev. A*, **25**, 978-989 (1982); *Science*, **225**, 983-989 (1984); Stillinger, F. H., *Science*, **267**, 1935-1939 (1995).

[51] Kirkpatrick, T. R., Thirumalai, D.,*Phys. Rev. B*, **36**, 5388-5397 (1987).

[52] Kirkpatrick, T. R., Wolynes, P. G.,*Phys. Rev. B*, **36**, 8552-8564 (1987).

[53] Kirkpatrick, T. R., Thirumalai, D., and Wolynes, P. G. *Phys. Rev. A*, **40**, 1045-1054 (1989).

[54] Mezard, M. and Parisi, G., *Phys. Rev. Lett.*, **82**, 747-750 (1999); Mezard, M. and Parisi, G., *J. Chem. Phys.*, **111**, 1076-1095 (1999).

[55] Coluzzi, B., Mezard, M., Parisi, G. and Verrocchio, P., *J. Chem. Phys.*, **111**, 9039-9052 (1999); Coluzzi, B., Parisi, G. and Verrocchio, P., *J. Chem. Phys.*, **112** 2933-2944 (2000); Coluzzi, B., Parisi, G. and Verrocchio, P., *Phys. Rev. Lett.*, **84**, 306-309 (2000).

[56] Cardenas, M., Franz, S. and Parisi, G., *J. Phys. A: Math. Gen.*, **31**, L163-L169 (1998).

[57] Stillinger, F. H., *J. Chem. Phys.*, **88**, 7818-7825 (1988).

[58] Speedy, R. J., *Mol. Phys.*, **95** 169-178 (1998).

[59] Scala, A., Starr, F. W., La Nave, E., Sciortino, F. and Stanley, H. E., *Nature*, **406**, 166-169 (2000).

[60] Sastry, S., *Phys. Rev. Lett.* **85**, 590-593 (2000).

[61] Sastry, S., *Nature*, **409**, 164-167 (2001).

[62] Sciortino, F., Kob, W. and Tartaglia, P., *Phys. Rev. Lett.*, **83**, 3214-3217 (1999).

[63] Buechner, S. and Heuer, A., *Phys. Rev. E*, **60**, 6507-6518 (1999); *Phys. Rev. Lett.*, **84**, 2168-2171 (2000).

[64] Starr, F. W., Sastry, S., La Nave, E., Scala, A., Stanley, H. E. and Sciortino, F., *Phys. Rev. E*, **63**, 041201-1-041201-10 (2001).

[65] Speedy, R. J., *J. Molec. Struct.*, **485-486**, 537-543 (1999).

[66] Sastry, S., *J. Phys. Cond. Mat.*, **12**, 6515-6523 (2000).

[67] Sastry, S. *Proceedings of Slow Dynamics and Freezing in Condensed Matter Systems, J. Nehru University, 2000. To appear in Phase Transitions.* (http://arXiv.org/abs/cond-mat/0101079).

[68] Sciortino, F. and Tartaglia, P., *Phys. Rev. Lett.*, **86**, 107-110 (2001).

[69] Speedy, R. J. and Debenedetti, P. G., *Mol. Phys.*, **88** 1293-1316 (1996).

[70] Crisanti, A. and Ritort, F., *Europhysics Lett.*, **51**, 147-153 (2000).

[71] Heuer, A. and Buechner, S., *J. Phys. Cond. Mat.*, **12**, 6535–6543 (2000).

[72] Saika-Voivod, I., Poole, P. H. and Sciortino, F., *Nature*, **412**, 514-517 (2001).

[73] Martinez, L.-M. and Angell, C. A., *Nature*, **410**, 663-667 (2001).

[74] Sciortino, F., La Nave, E., Tartaglia, P., *Phys. Rev. Lett.* **91**, 155701-1 – 155701-4 (2003).

[75] Saksaengwijit, A. Reinisch, J. and Heuer, A., *Phys. Rev. Lett.* bf 93 235701 (2004).

[76] Sciortino, F. Lecture notes for the School on "Unifying Concepts in Glass Physics III" Bangalore, June 2004, *J. Stat. Mech.* P05015 (2005).

[77] Wales, D.J. "Energy Landscapes" (Cambridge University Press, Cambridge, UK, 2003).

[78] Angelani, L., Di Leonardo, R., Ruocco, G., Scala, A. and Sciortino, F., *Phys. Rev. Lett.*, **85**, 5356-5359 (2000).

[79] Broderix, K., Bhattacharya, K. K., Cavagna, A., Zippelius, A. and Giardina, I., *Phys. Rev. Lett.*, **85**, 5360-5363 (2000).

[80] Keyes, T., *Phys. Rev. E*, **62**, 7905-7908 (2000).

[81] La Nave, E., Scala, A., Starr, F. W., Sciortino, F. and Stanley, H. E., *Phys. Rev. Lett.*, **84**, 4605-4608 (2000).

[82] Cavagna, A. http://arXiv.org/abs/0903.4264

[83] Doliwa, B. and Heuer, A., *Phys. Rev. E* **67** 031506 (2003); *Phys. Rev. Lett.* **91** 235501 (2003); *Phys. Rev. E* **67** 030501 (R) (2003).

[84] Denny, R.A., Reichman, D.R. and Bouchaud, J.-P., *Phys. Rev. Lett.* **90** 025503 (2003).

[85] Heuer, A. *J. Phys. Cond. Mat.* **20** 373101 (2008).

[86] Balucani, U. and Zoppi, M. "Dynamics of the Liquid State", (Oxford University Press, 1994).

[87] Reichman, D. Lecture notes for the School on "Unifying Concepts in Glass Physics III" Bangalore, June 2004, *J. Stat. Mech.* P05013 (2005).

[88] Bengtzelius, U., Götze, W. and Sjölander, A., *J. Phys. C: Solid State Phys.* **17** 5915–5934 (1984).

[89] Leutheusser, E.,*Phys. Rev. A* **29** 2765 - 2773, (1984).

[90] Kob, W., in *Slow relaxations and nonequilibrium dynamics in condensed matter*, J.-L. Barrat, M. Feigelman, J. Kurchan and J. Dalibard, eds. (Springer, Berlin, 2003) (available also at http://www.arXiv.org/cond-mat/0212344).

[91] Das, S. P. *Rev. Mod. Phys.* **76** 785 - 851 (2004).

[92] Götze, W., "Complex Dynamics of Glass-Forming Liquids: A Mode-Coupling Theory" (Oxford University Press, 2009).

[93] Biroli, G., Bouchaud, J.-P., Miyazaki, K. and Reichman, D. R. *Phys. Rev. Lett.*, **97**, 195701-1 – 195701-4 (2006).

[94] Berthier, L,. *Phys. Rev. Lett.* **91** 055701-1 – 055701-4 (2003).

[95] Berthier, L. *et al, Science*, **310** 1797 – 1800 (2005).

[96] Dalle-Ferrier, C. *et al, Phys. Rev. E* **76** 041510-1 – 041510-15 (2007).

[97] Capaccioli, S., Ruocco, G. and Zamponi, F., *J. Phys. Chem. B* 112:10652 - 10658 (2008).

[98] Berthier, L., Biroli, G., Bouchaud, J.-P., Kob, W., Miyazaki, K. and Reichman, D. R. *J. Chem. Phys.* **126** 184503-1 – 184503-21 (2007).

[99] Ritort, F. and Sollich, P., *Adv. Phys.*, **52** 219 – 342 (2003).

[100] Whitelam, S., Berthier, L. and Garrahan, J. P.*Phys. Rev. Lett.* **92**, 185705-1 – 185705-4 (2005).

[101] Cavagna, A., Lecture notes for the School on "Unifying Concepts in Glass Physics III" Bangalore, June 2004, *J. Stat. Mech.* P05012 (2005).

[102] Parisi, G., in *Slow relaxations and nonequilibrium dynamics in condensed matter*, J.-L. Barrat, M. Feigelman, J. Kurchan and J. Dalibard, eds. (Springer, Berlin, 2003) (available also at http://www.arXiv.org/cond-mat/0301157)

[103] Lubchenko, V., Wolynes, P. G., *Annu. Rev. Phys. Chem.* **58**, 235266 (2007).

[104] Goldstein, M., *J. Chem. Phys.*, **51**, 3728–3739 (1969).

[105] Schröder, T. B., Sastry, S., Dyre, J. C. and Glotzer, S. C., *J. Chem. Phys.*, **112**, 9834-9840 (2000).

[106] Bouchaud, J.-P. and Biroli, G., *J. Chem. Phys.* **121** 7347 – 7354 (2004).

[107] Franz, S., *J. Stat. Mech.* P04001 (2005).

[108] Franz, S. and Montanari, A., *J. Phys. A: Math. Theor.* **40**, F251 – F257 (2007).

[109] Bhattacharyya, S. M., Bagchi, B. and Wolynes, P. G. *Proc. Natl. Acad. Sci. USA* 105:16077 – 16082 (2008).

[110] Stevenson, J. D., Schmalian, J. and Wolynes, P. G. *Nature Physics* **2**, 268 – 274 (2006).

[111] Biroli, G., Bouchaud, J.-P., Cavagna, A., Grigera, T. S. and Verrocchio, P. *Nature Physics* **4**, 771 - 775 (2008).

[112] Karmakar, S., Dasgupta C. and Sastry, S. *Proc. Natl. Acad. Sci. USA* **106** 3675 - 3679 (2009).

Chapter 4

Dilute Magnets

Deepak Kumar

We review some selected topics from the theory of critical phenomena of diluted magnets near their percolation threshold. The theoretical interest in these studies lies in the circumstance that, for magnetic transitions near the percolation threshold, there is new and interesting physics due to the interplay of geometrical fluctuations and thermal fluctuations, both of which become very large at this point. A further stimulus to these studies has come from neutron scattering studies of some ideal magnetic substances, diluted to concentrations close to the percolation concentration [1–5]. The experimental work has provided rich information on the detailed nature of correlations in these systems, which is eminently suitable for comparison with the theoretical work on magnetic models.

4.1 Introduction

We begin with a short discussion of the relevant concepts of percolation theory in Section 4.2. In the next section, we introduce the problem of dilute magnets and discuss a simple-minded scaling approach, which serves as a useful background for digesting the experimental results and the more detailed treatments. In Section 4.4, we make a digression to discuss Pott's model and its relation to the percolation problem. This digression is needed to understand the formulation of the dilute Ising model by the replica method, which is presented in Section 4.5. This section also describes the renormalisation group calculation of the cross-over exponent at the percolation point in $6-\epsilon$ dimensions. Finally, in Section 4.6, we describe the salient features of experimental results and some simple, heuristic arguments to understand them.

4.2 Percolation Processes

The basic ideas of the percolation process have been discussed in a number of reviews [6–8]. Here we confine ourselves to introducing some important notions that are specifically required for our later arguments. Consider a lattice of N sites, in which each site can be in two states: occupied or empty. The sites are independent of each other. Let the probability of a site being occupied be p. Now a few basic quantities can be defined.

Cluster: A group of occupied sites connected by nearest neighbour bonds.

Mean cluster number: Let N_s denote the number of clusters of size s in a given configuration of the above process. Then the average value (over various configurations) of N_s/N is denoted by n_s and is called mean cluster number per site. n_s is basically the probability of having a cluster of size s, and it obviously depends upon p. For example, for a square lattice

$$n_1 = p(1-p)^4$$

$$n_2 = 2p^2(1-p)^6 \text{ etc.}$$

Percolation Transition When p is small, most occupied sites are isolated with only a few pairs and triplets. As p is increased, larger clusters are formed with increasing probability. When p is close to 1, most of the sites belong to a large cluster, extending from one end to the other. If we look at the lattice configuration as p changes, at one point a qualitative change occurs, namely the appearance of a large cluster spanning a large portion of the lattice. This notion is somewhat inexact for a finite lattice. But imagine an infinite lattice, where we can have an infinite cluster, which percolates through the system. As we raise p, the infinite cluster makes its appearance at a sharp concentration p_c, called the percolation threshold.

Percolation Probability P_∞ (p) More quantitatively, we define a quantity called percolation probability $P_\infty(p)$, as the fraction of occupied sites belonging to the infinite percolating cluster. Then $P_\infty(p) = 0$ if $p \langle p_c$ and $P_\infty(p) \neq 0$ if $p \geq p_c$, becoming unity as $p \to 1$. For the percolation transition, $P_\infty(p)$ plays the role of an order parameter.

A simple, useful relation can be derived by noting that the total number of sites can be divided amongst the following groups: (i) unoccupied, (ii) belonging to infinite cluster, (iii) belonging to one of the finite cluster. Thus

$$N = Np + N(1-p) \tag{4.1}$$

$$Np = N_\infty + {\sum_s}' sN_s \tag{4.2}$$

where N_∞ is the number of sites in the large cluster (the infinite cluster in the limit $N \to \infty$), and the prime in Eq. (4.2) implies that the summation includes

4.2. Percolation Processes

only finite clusters. Noting that $P_\infty(p) = N_\infty/Np$, one obtains by dividing Eq. (4.1) by N,

$$p = pP_\infty + {\sum_s}' sn_s \qquad (4.3)$$

In very analogous terms, one can also talk about the so-called 'bond-percolation process'. Here, the bonds of the lattice can be in two states-occupied or empty, with probabilities p and $1-p$ respectively. Occupied bonds connected to each other define a cluster. A percolation transition at a certain value of p can likewise occur in an infinite lattice. Like thermodynamic phase transitions, the percolation transition can also be described in terms of exponents, which characterize the singular behaviour of certain functions describing the phenomenon. Let us first consider the case of $P_\infty(p)$

$$P_\infty(p) = 0 \quad p < p_c \qquad (4.4)$$

$$P_\infty(p) \simeq (p - p_c)^{\beta_p} \quad p \geq p_c \qquad (4.5)$$

Then Eq. (4.3) implies that the singular part of $\sum' sn_s$ is

$$\left[{\sum_s}' sn_s\right]_{sing} \simeq (p - p_c)^{\beta_p} \qquad (4.6)$$

Next consider the mean square size of the finite cluster $\sum' s^2 n_s$. Recalling that n_s is the probability distribution for the cluster size, the mean square size is a measure of the fluctuation or the width of the cluster size distribution. This quantity is basically small both when $p \approx 0$ and $p \approx 1$. It is expected to become large around the transition point, $p \approx p_c$. At $p = p_c$, when the infinite cluster is just formed or about to be formed, this quantity is expected to diverge, as it is the thermodynamic analog of susceptibility or specific heat.

$$\left[{\sum}' s^2 n_s\right]_{sing} \propto |p - p_c|^{-\gamma_p} \qquad (4.7)$$

The analogy to magnetic phase transitions can be made quite formal, by defining an analog of free energy $f(p, h)$, per site as

$$f(p, h) = {\sum_s}' n_s e^{sh} \qquad (4.8)$$

Here h is an hypothetical field. Then the quantities equivalent to magnetisation, susceptibility etc. can be derived by differentiation with respect to h:

$$\langle s \rangle = \frac{\partial}{\partial h} f(p, h)|_{h=0} = {\sum_s}' sn_s = p(1 - P_\infty(p)) \qquad (4.9)$$

$$\langle s^2 \rangle = \frac{\partial^2}{\partial h^2} f(p, h)|_{h=0} = {\sum_s}' s^2 n_s \qquad (4.10)$$

One can also define pair-connectedness, which is the analog of the correlation function, by the equation: $C_{ij} = 1$ if the sites i and j are occupied and belong to the same cluster. $C_{ij} = 0$ otherwise.

Then for large r, r being the distance between the sites i and j, the quantity, $G_p(r) = \langle C_{ij} \rangle - P_\infty^2$, behaves as

$$G_p(r) \simeq \frac{e^{-r/\xi_p}}{r^{d-2+\eta_p}} \qquad (4.11)$$

where ξ_p is called the percolation correlation length. At $p = p_c$, ξ_p diverges as

$$\xi_p \simeq |p - p_c|^{-\nu_p} \qquad (4.12)$$

In parallel to the thermodynamic critical phenomena [9], one introduces a scaling ansatz for $f(p,h)$. It is based on a central idea due to Kadanoff [10] and holds near the percolation threshold, where the correlation length is large. The idea is that we divide the system in blocks of sites and define new degrees of freedom associated with the blocks by clumping together all the microscopic degrees of freedom of the sites within the block. When the linear size λ of the block is smaller than ξ_p, the key assumption is that the physics of the problem in terms of the new degrees of freedom is the same, as that in terms of the original degrees of freedom, apart from a change of scale in the basic parameters of the system. In the present context, the content of this idea can be expressed by the following equation,

$$N f(\delta p, h) = \frac{N}{\lambda^d} f(\delta p', h') \qquad (4.13)$$

where δp denotes $p - p_c$, d denotes the dimension of the system, and p' and h' are the scaled variables associated with the blocks of size λ. Note that the number of new degrees of freedom is N/λ^d and the block free energy has the same dependence on the scaled parameters as the original free energy has on p and h. The block variables are assumed to be related to the original variables by the following equations.

$$\delta p' = p' - p_c = \lambda^a \delta p \qquad (4.14)$$
$$h' = \lambda^b h \qquad (4.15)$$

This leads to the generalized homogeneity hypothesis for the free energy

$$f(\delta p, h) = \lambda^{-d} f(\lambda^a \delta p, \lambda^b h) \qquad (4.16)$$
$$= (\delta p)^{d/a} f\left(1, h/(\delta p)^\Delta\right) \qquad (4.17)$$

where in the second line we have chosen $\lambda = (\delta p)^{-1/a}$ The further physical connection is made by taking $\lambda = \xi_p$ and $a = \nu_p^{-1}$. One can also define the analog of specific heat index α, by the relation $d\nu_p = 2 - \alpha_p$. The singular behaviour of $f(p,h)$ and relations between the various exponents defined above follow from the form,

$$f(\delta p, h) = (\delta p)^{2-\alpha_p} f\left(1, h/(\delta p)^\Delta\right) \qquad (4.18)$$

4.3. Dilute Magnets: Scaling Theory

It is easily verified from the above relations that the power law exponents obey the following relations.

$$\alpha_p + 2\beta_p + \gamma_p = 2 \quad (4.19)$$

The analogs of other exponent relations obtained in the thermodynamic critical phenomena are similarly obtained from the basic relation Eq. (4.18). The Kadanoff's block procedure can now be done quite formally by the Renormalization Group (RG) method due to Wilson [11]. The RG method also allows us to calculate the exponents a, b etc. in a systematic manner.

4.3 Dilute Magnets: Scaling Theory

A typical model used to discuss the essential physics of insulating dilute magnets is the Heisenberg model given by the following modified Hamiltonian

$$H = -J \sum_{ij} p_{ij} \vec{S}_i . \vec{S}_i \quad (4.20)$$

where the summation runs over nearest neighbour bonds only; $p_{ij} = 1$ with probability p and $p_{ij} = 0$ with probability $1 - p$. Thus a realization of the probability distribution of p_{ij}'s tells us which bonds are present or absent in a given configuration. If the vector spins $\vec{S}_i$ are replaced by Ising variables σ_i, we have the diluted Ising model.

As p, the average concentration of bonds, decreases from 1, the transition temperature T_c decreases as the averaged molecular field experienced by a spin decreases with the concentration. It is quite simple to see that $T_c(p)$ should become zero at or somewhat above p_c. This is so because below p_c, the system is divided into isolated clusters of finite size, and the finite sized clusters cannot support any long-range order. The various calculations [7, 12, 13] have now established the fact that $T_c(p) \to 0$ at $p = p_c$. The typical phase diagram for a dilute magnet is shown in Fig. (4.1). The point $(p = p_c, T = 0)$ is a multi-critical point, in the sense that at this point, both the thermodynamic correlation length ξ_T and the percolation correlation length ξ_p diverge [14]. One of the purposes of these chapter is to present calculations and arguments which elucidate the behaviour in this region.

Let us begin by considering the equation of state near $T = 0$. All spins within a clusters are parallel, so that the effective moment of the s-cluster is μs, where μ is the moment of a single spin. Thus at small fields, the magnetization, M is,

$$M = {\sum_s}' \mu N n_s s \tanh \frac{\mu s H}{kT} + \mu N p P_\infty(p) \quad (4.21)$$

where the first term gives the magnetisation of the finite clusters, while the second term represents the spontaneous magnetisation in the infinite cluster. The susceptibility χ is given by

$$\chi = \frac{\partial M}{\partial H}|_{H \to 0} = \frac{N \mu^2}{kT} {\sum}' s^2 n_s \quad (4.22)$$

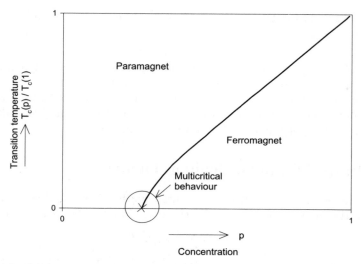

Figure 4.1: Schematic variation of the critical temperature $T_c(p)$ of a dilute ferromagnet with the concentration p, p_c denoting the percolation threshold.

or

$$\chi_p = \frac{\chi kT}{N\mu^2} = \langle s^2 \rangle \simeq |p - p_c|^{-\gamma_p} \qquad (4.23)$$

Eq. (4.23) tells us that the magnetic susceptibility near $T = 0$, is proportional to mean square cluster size, which is the susceptibility for the percolation transition and diverges as p approaches the percolation threshold.

At this point one asks, as to what happens at higher temperatures, in particular, how does the divergence occur when the multicritical point $(p - p_c, T = 0)$ is approached along the temperature axis. A reasonable ansatz based on our experience with scaling theories is that χT is a scaling function of $|p-p_c|$ and some field g depending upon temperature [14]. Thus

$$\frac{\chi kT}{N\mu^2} = |p - p_c|^{-\gamma_p} \Psi\left(\frac{g(T)}{|p - p_c|^\phi}\right) \qquad (4.24)$$

the index ϕ is called the cross-over index and it characterizes the competition between thermal and geometrical fluctuations. This form can also be derived from Kadanoff's block picture. One can write

$$\chi_p(\delta p, T) = \lambda^c G(\delta p \lambda^a, g(T)\lambda^b) \qquad (4.25)$$

where the rhs gives the susceptibility in terms of the scaled variables of the blocks of linear size λ. Now we choose $\lambda = \xi_p = (\delta p)^{-\nu_p}$ and recall that $\nu = 1/a$. Then identifying $c\nu_p = \gamma_p$, one obtains

$$\frac{\chi kT}{N\mu^2} = (\delta p)^{-\gamma_p} G\left(1, \frac{g(T)}{(\delta p)^{b\nu_p}}\right) \qquad (4.26)$$

4.3. Dilute Magnets: Scaling Theory

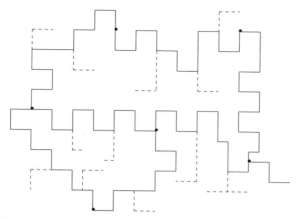

Figure 4.2: The node-link model of the infinite percolation cluster. The two lengths ξ_p and L which characterize the model are indicated. The dotted line show the dangling ends

Basically at the percolation point, both the thermal and the geometrical correlation lengths diverge. One expects that various thermodynamic functions are determined by the ratio of these two lengths. This allows us to identify $1/b = \nu_T$ as the exponent for the divergence of the thermal correlation length, ie $\xi_T = g(T)^{-\nu_T}$, then $\phi = \nu_p/\nu_T$.

Similar scaling relations can be written for the free energy, specific heat etc, e.g.

$$\frac{C_v}{Nk} = |\delta p|^{-\alpha_p} H\left(\frac{g(T)}{(\delta p)^\phi}\right) \quad (4.27)$$

In order to describe the critical behaviour away from the multi-critical point $\Psi(x)$ is assumed to have a singularity of the form [14]

$$\Psi(x) \propto A(x - x_c)^{-\gamma} \quad (4.28)$$

Then

$$\frac{\chi kT}{N\mu^2} \simeq A|g - g_c(p)|^{-\gamma}|p - p_c|^{\gamma\phi - \gamma_p} \quad (4.29)$$

Next we present a simple scaling argument due to Lubensky [15] which determines $g(T)$ and ϕ. This argument is based on a heuristic picture of the infinite cluster called the node-link model, due to Skal and Shklovskii [16,17]. In this picture, all the dangling bonds from the infinite cluster are removed and the infinite cluster is viewed as a superlattice of nodes, defined as points where there are more than two disjoint paths to infinity. This idea is illustrated in Fig. 4.2. The nodes are called adjacent if the path connecting them does not pass through another node. There are two lengths in the picture. First is the average real distance (measured in terms of lattice distance) between the nodes, which is of the order of correlation length ξ_p. Second is the average number of bonds between the adjacent nodes-or the length L of the crooked path joining the adjacent nodes. Clearly $L > \xi_p$. Some authors [18] have taken the paths to be self-avoiding walks, in which event $\xi_p \simeq L^{\nu_{SAW}}$. But clearly, the self-avoiding walk represents an upper limit for L, so that [15]

$$\xi_p < L \leq \xi_p^{1/\nu_{SAW}} \tag{4.30}$$

As p approaches p_c from above, L diverges like $(p-p_c)^{-\zeta}$. In view of Eq. (4.23)

$$\nu_p \leq \zeta \leq \nu_p/\nu_{SAW} \tag{4.31}$$

Now let us consider the magnetic correlations. The correlations between spins with separation less than L bonds, are essentially one-dimensional in character. If we are at a temperature T, such that the one-dimensional correlation length $\xi_1(T)$ is less than L, the system behaves like a collection of one dimensional segments. On the other hand, if T is such that $\xi_1(T) > L$, the true d–dimensional nature of the network is felt and the ferromagnetic ordering can occur. Thus the competition between the geometrical and thermal fluctuations can be described in a simple minded way by $L/\xi_1(T)$, the ratio of the two lengths [15]. In the spirit of the scaling theory, this ratio may be taken to be a scaling variable, to write the scaling form for susceptibility as

$$\frac{\chi kT}{N\mu^2} = |p - p_c|^{-\gamma_p} \Psi\left(\frac{\xi_1^{-1}(T)}{|p-p_c|^\zeta}\right) \tag{4.32}$$

According to this prescription, the cross-over index ϕ is determined from purely geometrical considerations to be equal to ζ. The temperature scaling variable $g(T) = \xi_1^{-1}(T)$. The quantity $\xi_1^{-1}(T)$ is well known. At low temperatures

$$\xi_1^{-1}(T) = e^{-2J/kT} \quad \text{for Ising system} \tag{4.33}$$

$$\xi_1^{-1}(T) = \frac{m-1}{2m}\frac{kT}{J} \quad \text{for m-component classical spins} \tag{4.34}$$

Using Eq. (4.29), one can write the form of the critical line as it rises from the percolation point. One has

$$T_c(p) = \frac{2J}{\zeta|\ln(p-p_c)|} \quad \text{for Ising spins} \tag{4.35}$$

$$T_c(p) = x_c(p-p_c)^\zeta \quad \text{for Heisenberg spins} \tag{4.36}$$

The present theoretical status of the problem is briefly as follows. For the two dimensional Ising model, there is an exact calculation [19] which verifies Eq. (4.35), and gives in addition $\phi = 1$. For the Ising system the renormalisation group calculations [20], to be described later, give $\phi = 1$ in $6-\epsilon$ dimensions, to all orders of ϵ–expansion perturbation theory [21]. For the Heisenberg model, the problem has been treated by approximate real space renormalisation group

4.4. Potts Model and Percolation

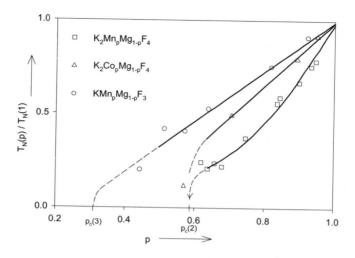

Figure 4.3: Experimental data on the variation of Neel temperatures on three compounds: $K_2(MnMg)F_4$ (2-dimensional Heisenberg Antiferromagnet); $K_2(CoMg)F_4$ (2-dimensional Ising Antiferromagnet); $K(MnMg)F_3$ (3- dimensional Heisenberg Antiferromagnet). The shape of the experimental curves is in general accord with predictions for Ising and Heisenberg behaviour, as in Eqs. (4.35) and (4.36).

methods [22–24]. Most of these calculations verify Eq. (4.36) in general and give $\phi \simeq 1$ for $d = 3$ and $\phi \simeq 1.2$ to 1.4 for $d = 2$.

Experimentally also one finds the distinct difference between the Ising like and the Heisenberg like systems. In Fig. 4.3, we show the data [25] on the variation of Neel temperature of three compounds, which are rather faithful realisations of Ising and Heisenberg models. $K_2(MnMg)F_4$ is a two demensional Heisenberg anti-ferromagnet with a slight anisotropy, $K_2(CoMg)F_4$ is a two dimensional Ising antiferromagnet and $K(MnMg)F_3$ is a three dimensional Heisenberg antiferromagnet with a weak anisotropy. The Heisenberg systems show the cross-over to the Ising behaviour due to small anisotropies at low temperatures. Otherwise the behaviour is well in accord with Eqs. (4.35) and (4.36).

4.4 Potts Model and Percolation

In a very interesting paper, Fortuin and Kasteleyn [26] showed that the percolation problem can be treated as a regular statistical mechanical problem in terms of Pott's model. This section is devoted to describing this connection [26, 27]. Pott's model is a generalisation of the Ising model. At each lattice site i, one has a variable σ_i which can be in any of the m states. The Hamiltonian of the system is written as

$$-\beta H = \sum_{\langle ij \rangle} mK(\delta_{\sigma_i,\sigma_j} - 1) + mH \sum_i (\delta_{\sigma_i,1} - 1) \quad (4.37)$$

where the first summation is over all the nearest neighbour pairs. The partition function Z is:

$$Z = Tr_{\{\sigma_i\}} \exp\left[\sum_{\langle ij \rangle} mK(\delta_{\sigma_i,\sigma_j} - 1) + mH \sum_i (\delta_{\sigma_i,1} - 1)\right] \quad (4.38)$$

This expression can be simplified by noting that

$$e^{mK(\delta_{\sigma_i,\sigma_j}-1)} = e^{-mK} + (1 - e^{-mK})\delta_{\sigma_i,\sigma_j} \quad (4.39)$$

Now define

$$p = 1 - e^{-mK} \quad (4.40)$$

Then Z can be written as

$$Z = \sum_{\{\sigma_i\}} \prod_{ij} \left[1 - p + p\delta_{\sigma_i,\sigma_j}\right] e^{mH \sum_i (\delta_{i,1}-1)} \quad (4.41)$$

$$Z = (1-p)^{N_B} \sum_{\sigma_i} \left[1 + \frac{p}{1-p} \sum_{\langle ij \rangle} \delta_{\sigma_i,\sigma_j} \right.$$

$$\left. + \left(\frac{p}{1-p}\right)^2 \sum_{\langle ij \rangle} \sum_{\langle hl \rangle}{}' \delta_{\sigma_i,\sigma_j}\delta_{\sigma_h,\sigma_1} + \ldots \right] e^{mH \sum_i (\delta_{i,1}-1)} \quad (4.42)$$

where N_B is the total number of nearest neighbour bond. $N_B = Nz/2$, where z is the number of nearest neighbours per site. Now each term of the expansion is given a graphical representation. Draw N−lattice points. For each factor $(\frac{p}{1-p})\delta_{\sigma_i,\sigma_j}$, draw a bond between sites i and j. There can be bonds only between nearest neighbour sites, and there can be only one bond between two sites. It is easy to see that there is one to one correspondence between the terms of the series in Eq. 4.43 and such N−point graphs. A typical graph is shown in Fig. (4.4)(a). Now Z may be written as

$$Z = \sum_g (1-p)^{N_B} \left(\frac{p}{1-p}\right)^{N_B(g)} \prod_{\text{product over clusters}} C_{cluster} \quad (4.43)$$

where the summation is over all possible graphs g, $N_B(g)$ denotes the number of bonds in the graph g, and $C_{cluster}$ denotes the contribution of the cluster. Let $A_s(H)$ denote the contribution of the s-cluster and $Nn(s,g)$ denote the number

4.4. Potts Model and Percolation

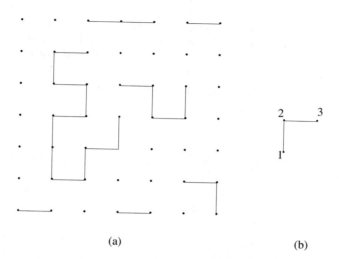

Figure 4.4: (a) Graphical representation of a typical term occurring in the expansion of the partition function of the Potts model as given in Eq. (4.43). (b) Cluster whose contribution is calculated in Eq. (4.45).

of clusters containing s sites in the graph g. The factor $p^{N_B(g)}(1-p)^{N_B-N_B(g)}$, is the probability of occurrence of the graph g and is denoted by $P(g)$. Then,

$$Z = \sum_g P(g) \prod_s [A_s(H)]^{Nn(s,g)} \qquad (4.44)$$

The cluster contribution $A_s(H)$ is easily evaluated. For illustration, consider the graph of Fig. 4.4(a). The contribution is,

$$\sum_{\sigma_1,\sigma_2,\sigma_3} \delta_{\sigma_1,\sigma_2} \delta_{\sigma_2,\sigma_3} \; e^{(\delta_{\sigma_1,1}+\delta_{\sigma_2,1}+\delta_{\sigma_3,1}-3)mH}$$

$$= \sum_{\sigma_1} e^{3(\delta_{\sigma_1,1}-1)mH}$$

$$= 1 + (m-1)e^{-3mH} \qquad (4.45)$$

The generalisation to an arbitrary cluster yields

$$A_s(H) = 1 + (m-1)e^{-mHs} \qquad (4.46)$$

Thus

$$Z = \sum P(g) \prod_s [1 + (m-1)e^{-smH}]^{Nn(s,g)} \qquad (4.47)$$

For $H = 0$, one can express the result as

$$Z = \langle m^{Nn(s,g)} \rangle \qquad (4.48)$$

where the averaging is over the graphs with the weight factor $P(g)$. Now consider the limit $m \to 1$, which is unphysical from the point of view of statistical mechanics. If we set $m = 1$ in Eq. (4.48), Z becomes 1. However, consider

$$f_1 = \lim_{N \to \infty} \lim_{m \to 1} \frac{1}{N(m-1)} \ln Z \tag{4.49}$$

One may rewrite f_1 as

$$f_1 = \lim_{N \to \infty} \frac{1}{N} \frac{\partial}{\partial m}(\ln Z)|_{m=1} = {\sum_s}' \left\langle n(s,g) \frac{\partial}{\partial m} \ln[1 + (m-1)e^{-smH}]|_{m=1} \right\rangle_g$$

$$= \sum_g P(g) {\sum_s}' n(s,g) e^{-sH} = {\sum_s}' n_s e^{-sH} \tag{4.50}$$

Comparing Eq. (4.50) with Eq. (4.8), we find that f_1 reduces to the percolation analog of free energy, provided one makes the correspondence that the bond occupation probability $p = 1 - e^{-K}$, where K is the coupling parameter in the Pott's Hamiltonian. This correspondence is very useful since it allows the use of the machinery of statistical mechanics for the percolation problem.

4.5 Diluted Ising Model

This section describes a calculation due to Stephen and Grest [20] on the critical properties of the diluted Ising model near its percolation threshold. The method employs the replica trick to obtain an effective non-random Hamiltonian, which at $T = 0$ reduces to an effective Hamiltonian for the percolation problem. The renormalisation group method is then used to characterise the singular behaviour at the percolation point.

The partition function Z for the diluted Ising model is:

$$Z = Tr_{\{\sigma_i\}} \exp\left[\beta J \sum_{ij} p_{ij}(\sigma_i \sigma_j - 1) + \beta H \sum_i \sigma_i\right] \tag{4.51}$$

where p_{ij} are the random variables defined after Eq. (4.20). For calculating the configurationally averaged free energy, one employs the replica method. Here one first calculates Z^n.

$$Z^n = \sum_{\{\sigma_i^\alpha\}} \exp\left[\sum_{\alpha=1}^n \left\{ K \sum_{\langle ij \rangle} p_{ij}(\sigma_i^\alpha \sigma_j^\alpha - 1) + \sum_i h\sigma_i^\alpha \right\}\right]$$

where $K = \beta J$. For the present we consider the $h = 0$ case. Averaging over p_{ij}'s yields,

$$\langle Z^n \rangle = \sum_{\{\sigma_i^\alpha\}} \prod_{\langle ij \rangle} \left[1 - p + p e^{K \sum_{\alpha=1}^n (\sigma_i^\alpha \sigma_j^\alpha - 1)} \right] \tag{4.52}$$

4.5. Diluted Ising Model

The average free energy $\langle F \rangle = -kT\langle \log Z \rangle$ is obtained from the following relation:

$$\langle F \rangle = -kT\frac{\partial}{\partial n}\langle Z^n \rangle|_{n=0} \tag{4.53}$$

The quantity $\langle F^n \rangle$ is to be calculated from an effective Hamiltonian given by

$$\beta H_{eff}(n) = \sum_{\langle ij \rangle} \ln\left[1 + \nu e^{K\sum_\alpha (\sigma_i^\alpha \sigma_j^\alpha - 1)}\right] \tag{4.54}$$

where $\nu = p/1 - p$, and a constant is ignored in writing Eq. (4.54). To analyse this Hamiltonian, especially at low temperatures, it proves convenient to introduce projection operators. First consider,

$$p_\pm^\alpha(ij) = \frac{1 \pm \sigma_i^\alpha \sigma_j^\alpha}{2} \tag{4.55}$$

p_+^α is 1 when the spins are parallel, or when the bond is in lower energy state and zero otherwise. Likewise p_-^α is 1 when the bond is in higher energy state and zero otherwise. Using these projection operators, we define

$$P_0(ij) = \frac{1}{2^n}\prod_\alpha (1 + \sigma_i^\alpha \sigma_j^\alpha)$$

$$P_1(ij) = \frac{1}{2^n}\sum_\alpha (1 - \sigma_i^\alpha \sigma_j^\alpha)\prod_{\beta \neq \alpha}(1 + \sigma_i^\alpha \sigma_j^\alpha)$$

$$P_n(ij) = \frac{1}{2^n}\prod_\alpha (1 - \sigma_i^\alpha \sigma_j^\alpha) \tag{4.56}$$

The operator $P_k(ij)$ will have k factors of p_- operators and $n - k$ factors of p_+ operators in all possible combinations. Thus these operators project out states corresponding to different excitation energies of the ij- bond.

Now we express $H_{eff}(n)$ in terms of these projection operators

$$\beta H_{eff} = \sum_{k=0}^{n}\sum_{\langle ij \rangle} A_k P_k(ij) \tag{4.57}$$

A_k's can be calculated to be,

$$A_k = \ln(1 + \nu e^{-2kK}) \tag{4.58}$$

The advantage of this decomposition is that now each term $P_k(ij)$ corresponds to a definite excitation energy for the bond (ij). To discuss the cross-over at low temperatures, it is sufficient to keep the first two terms only. Here the expansion variable is clearly e^{-2K} and by keeping only the first two terms, we are ignoring terms of order e^{-4K} and lower. Thus

$$\beta H_{eff} = \sum_{\langle ij \rangle} \left[\ln(1+\nu) \prod_\alpha \frac{(1+\sigma_i^\alpha \sigma_j^\alpha)}{2} \right.$$
$$\left. + \ln(1+\nu e^{-2K}) \sum_\alpha \frac{1-\sigma_i^\alpha \sigma_j^\alpha}{2} \cdot \prod_{\beta(\neq\alpha)} \frac{1+\sigma_i^\beta \sigma_j^\beta}{2} \right] \quad (4.59)$$

First consider the $T=0$ limit. Only the first term of Eq. (4.59) survives. Now the partition function is:

$$e^{\beta_{eff} H_{eff}} = \prod_{\langle ij \rangle} \exp\left[\ln(1+\nu)\delta_{\sigma_i^\alpha,\sigma_j^\alpha}\right] = \prod_{\langle ij \rangle} \left[1+\nu\delta_{\sigma_i^\alpha,\sigma_j^\alpha}\right] \quad (4.60)$$

This is just the Pott's model viz., compare it with Eq. (4.41) with $h=0$. Since the collective variable $\{\sigma_i^\alpha\}$ take 2^n values, we have a 2^n Pott's model. Since in the replica method we have to take $n \to 0$ limit, Eq. (4.60) reduces to 1-state Pott's model, which is indeed the limit in which Pott's model describes the percolation problem.

In order to proceed further, we introduce the following site variables [20]

$$\tau_i^\alpha = \sigma_i^\alpha$$
$$\tau_i^{\alpha\beta} = \sigma_i^\alpha \sigma_i^\beta \quad \alpha \neq \beta$$
$$\tau_i^{\alpha\beta\gamma} = \sigma_i^\alpha \sigma_i^\beta \sigma_i^\gamma \quad \alpha \neq \beta \neq \gamma \quad (4.61)$$

In terms of these, one can define

$$\mu^{(1)}(i,j) = \sum_\alpha \tau_i^\alpha \tau_j^\alpha$$

$$\mu^{(2)}(i,j) = \sum_{(\alpha,\beta)} \tau_i^{\alpha\beta} \tau_j^{\alpha\beta}$$

$$\mu^n(i,j) = \tau_i^{\alpha\beta} \ldots \tau_j^{\alpha\beta} \ldots \quad (4.62)$$

Now the Hamiltonian (4.60) can be written as

$$\beta H_{eff} = -\sum_{ij} \sum_{p=1}^n K_p \mu^{(p)}(i,j) \quad (4.63)$$

with

$$K_p = \ln(1+\nu) - 2p\nu e^{-2K} \quad (4.64)$$

Thus we see that at finite temperature K_p's become different thereby breaking the symmetry of the Pott's model. The problem of treating the crossover from percolation to thermal behaviour becomes identical to the more familiar anisotropy crossover in the vector spin models.

The application of the renormalisation group methods to the effective Hamiltonian of Eq. (4.63) in $6-\epsilon$ dimensions proceeds along the usual lines [20] which we shall not describe in detail. First a Hubbard-Stratanovich transformation is used to go over to continuous variables $\psi_\alpha(i)$ corresponding to τ_i's. Then the lattice is replaced by a continuum, leading to an Hamiltonian of the form

$$\beta H_{eff} = \frac{1}{2} \int d\psi^\alpha(\vec{x}) \left[\sum_{(\alpha)} \left\{ r_{(\alpha)} \psi^2_{(\alpha)}(\vec{x}) + (\vec{\nabla} \psi_{(\alpha)})^2 \right. \right.$$
$$\left. \left. - \omega \sum \lambda_{(\alpha)(\beta)(\gamma)} \psi_{(\alpha)}(\vec{x}) \psi_{(\beta)}(\vec{x}) \psi_{(\gamma)}(\vec{x}) \right\} \right] \quad (4.65)$$

where $\lambda_{(\alpha)(\beta)(\gamma)}$ is unity if each replica appears twice, once in each of two different indices, or not at all, and zero otherwise. Further

$$r_\alpha = r_1 = r_0 - e^{-2K}$$

$$r_{\alpha\beta} = r_2 = r_0 - 2e^{-2K}$$

where

$$r_0 \simeq p_c - p \quad (4.66)$$

Leaving details to the references, here we merely quote that the cross-over exponent associated with e^{-2K} is found to be

$$\phi = 1 + \frac{2^n - 1}{10 - 3 \times 2^n} \epsilon \quad (4.67)$$

which becomes 1 as $n \to 0$. Wallace and Young [21] have used group theoretic arguments to deduce that $\phi = 1$ to all higher order in ϵ. Thus this calculation provides a detailed justification of the results based on simple scaling theory and obtains explicitly the value of ϕ.

4.6 Neutron Scattering Results and a Simple Model of Correlations

In this section we discuss some of the salient features of magnetic correlations in diluted systems near the percolation threshold, as deduced from the neutron scattering experiments [1–5]. The systems studied in detail by the Brookhaven laboratory group are: $Rb_2 Mn_p Mg_{1-p} F_4$ (2-dimensional Heisenberg antiferromagnet), $Rb_2 Co_p Mg_{1-p} F_4$ (2-dimensional Ising antiferromagnet) and $Mn Zn F_2$ (3-dimensional Heisenberg antiferromagnet). The Heisenberg systems mentioned above have a small amount of uniaxial anisotropy.

The neutron scattering experiments essentially measure the Fourier transforms of static longitudinal and transverse correlation functions, as defined below

$$S_{\|}(q) \sim \int e^{i\vec{q}\cdot\vec{r}} \langle S^z(\vec{0})S^z(\vec{r})\rangle d\vec{r} \qquad (4.68)$$

$$S_{\perp}(q) \sim \int e^{i\vec{q}\cdot\vec{r}} \langle S^x(\vec{0})S^x(\vec{r})\rangle d\vec{r} \qquad (4.69)$$

where the z-axis is taken along the direction in which the spontaneous magnetisation points. The main inferences drawn from these experiments are [1–5]:

(a) Both $S_{\|}(\vec{q})$ and $S_{\perp}(\vec{q})$ have Lorentzian forms i.e.,

$$S_{\|}(\bar{q}) = \frac{k_{\|}^{\eta}}{k_{\|}^2 + q^2} \qquad (4.70)$$

$$S_{\perp}(q) = \frac{k_{\perp}^{\eta}}{k_{\perp}^2 + q^2} \qquad (4.71)$$

where $\vec{q} = \vec{Q} - \vec{q}_0$, with $\vec{Q}$ denoting the momentum transfer and $\vec{q}_0$ the wave vector of the long-range order (e.g., $|\vec{q}_0| = \pi/a$ for antiferromagnetic ordering). The index η varies between 0.34 to 0.42, for $d = 2$ systems. This value of η is rather large considering that for the pure Ising system $\eta = 0.25$.

(b) The inverse correlation length k for both longitudinal and transverse correlations obeys the following additive property,

$$k = k_p + k_T \qquad (4.72)$$

where k_p is the inverse correlation length for percolation and k_T is the inverse of the thermal correlation length. Furthermore k_T can be expressed in terms of one-dimensional correlation length ξ_1 in the following way

$$k_T = (\xi_1)^{-\nu_T} \qquad (4.73)$$

As discussed earlier, Stanley et. al. [18] have taken this to be strongly suggestive of the fact that the magnetic correlations propagate along one-dimenional paths. Further, if one-dimensional paths are taken to be self-avoiding walks $\nu_T = \nu_{SAW}$. However, experimentally it is found that $\nu_T = 1.32$; $\phi = 1.03$ for $Rb_2\ Co/Mg\ F_4$ and $\nu_T = 0.91$ and $\phi = 1.51$ for $Rb_2\ Mn/Mg\ F_4$. This tells us that ν_T is not determined geometrically alone as suggested by self-avoiding walk picture, but does depend upon the nature of spins.

(c) Even a small anisotropy in $Rb_2\ Mn/Mg\ F_4$ plays a dominant role as the temperature is lowered. This is reflected in big difference in $S_{\|}$ and $S_{\perp}$. As temperature is lowered $\xi_{\perp}$ saturates while $\xi_{\|}$ continues increasing, but its form crosses over from Heisenberg type to Ising type.

4.6. Neutron Scattering Results and a Simple Model of Correlations

Kumar, Pandey and Barma [23] have given a detailed analysis of some of these points. Here we develop some of the simple consequences of the idea that magnetic correlations propagate along one-dimensional paths. Consider the correlation function $C_{||}(\vec{r})$.

$$C_{||}(\vec{r}) = \langle S_z(0) S_z(\vec{r})\rangle - \langle S_z\rangle^2 \tag{4.74}$$

Near $p \approx p_c$, the two distant sites are connected by one or few number of paths. This implies that magnetic correlations are mainly of one-dimensional character and are independent of the geometrical configuration of the cluster. Thus one may factorise $C_{||}(\vec{r})$ into two factors.

$$C_{||}(\vec{r}) = C'_M(\vec{0},\vec{r}) G_p(\vec{0},\vec{r}) \tag{4.75}$$

where $G_p(\vec{0},\vec{r})$ is the geometrical correlation function given in Eq. (4.11) and $C'_M(\vec{0},\vec{r})$ is the magnetic correlation function on a one-dimensional chain passing through $\vec{0}$ and $\vec{r}$. The correlation function for a magnetic chain is:

$$\langle S_0 S_n \rangle = u^n \tag{4.76}$$

$$\langle S_0 S_n \rangle = e^{-n/\xi_1} \tag{4.77}$$

where

$$u = \tanh \beta J \quad \text{for Ising model} \tag{4.78}$$

$$u = \coth \beta J - \frac{1}{\beta J} \quad \text{for the classical Heisenberg model} \tag{4.79}$$

and

$$\xi_1^{-1} = -\ln u \tag{4.80}$$

Let $p(n,\vec{r})$ give the relative probability that a self avoiding path going through $\vec{0}$ and $\vec{r}$ has n-steps. Then

$$C'_M(\vec{r}) = \sum_n p(n,\vec{r}) e^{-n/\xi_1} \tag{4.81}$$

To evaluate Eq. (4.81) one may use a rigorous result due to Fisher [29] which says that if $p(n,\vec{r})$ can be expressed in a scaling from of the type

$$p(n,\vec{r}) = R_n^{-d} \phi(r/R_n) \tag{4.82}$$

where

$$R_n = R_0 n^{\nu_T} \tag{4.83}$$

then

$$\sum_n p(n,\vec{r}) e^{-n/\xi_1} \simeq e^{-r/\xi_T} \tag{4.84}$$

with

$$\xi_T = \xi_1^{\nu_T} \tag{4.85}$$

Though the results quoted above are intended for self-avoiding walks, they apply as well to the paths envisaged here, which, have other manners of constraints in addition to the self-avoiding walk constraint. Finally we write

$$C'_M(\vec{r}) \sim \frac{e^{-r/\xi_T}}{r^{\eta_M}} \qquad (4.86)$$

where we have introduced an extra factor of $r^{-\eta_M}$. This is to take into account the fact that when $r \geq L$, the correlations spread two-dimensionally. Here we argue that the one-dimensional character of the exponential term is still maintained because the spins are still connected by only a few parallel paths. Combining Eq. (4.86) with Eq. (4.11) we obtain,

$$C_{||}(\vec{r}) = \frac{e^{-(\xi_p^{-1}+\xi_T^{-1})r}}{r^{\eta_M+\eta_p}} \qquad (4.87)$$

$$or C_{||}(\vec{r}) = \frac{e^{-r/\xi}}{r^{\eta}} \qquad (4.88)$$

with
$$\xi^{-1} = \xi_p^{-1} + \xi_T^{-1} \quad \text{and} \quad \eta = \eta_M + \eta_p \qquad (4.89)$$

The fourier transform of $C_{||}(\vec{r})$ for small q then yields

$$S_{||}(q) = \frac{k^{\eta}}{k^2 + q^2} \qquad (4.90)$$

Though, our arguments are inadequate to calculate η_M, they do suggest why η is so large. Since $\eta_p = 0.25$, one expects $\eta_M \simeq .10$ to $.17$. Further our simple argument explains the addition of the inverse correlation lengths in a very natural manner.

There are several other topics that have recently been analysed e.g., the calculations for vector spins, role of anisotropy, the detailed forms for scaling functions etc. But these are not included in the scope of this chapter, which was designed to give the reader just the flavour of the subject.

References

[1] R. J. Birgeneau, R. A. Cowley, G. Shirane and H. J. Guggenheim, Phys. Rev. Lett 37, 940 (1976).

[2] R. A. Cowley, G. Shirane, R. J. Birgeneau and H. J. Guggenheim, Phys. Rev. B15, 4292 (1977).

[3] R. A. Cowley, G. Shirane, R. J. Birgeneau and E. C. Svensson, Phys. Rev. Lett. 39, 894 (1977).

[4] R. J. Birgeneau, R. A. Cowley, G. Shirane, J. A. Tarvin and H. J. Guggenheim, Phys. Rev. B21, 317 (1980).

[5] R. A. Cowley, R. J. Birgeneau, G. Shirane, H. J. Guggenheim and H. Ikeda, Phys. Rev. B21, 4038 (1980).

[6] G.I. Menon and P. Ray, Chapter in *this volume*

[7] S. Kirkpatrick, Rev. Mod. Phys. 45, 570 (1973).

[8] D. Stauffer, *Introduction to Percolation Theory*, (Taylor and Francis, London, 1985).

[9] S. K. Ma, *Modern Theory of Critical Phenomena*, (Benjamin, New York, 1975)

[10] L. P. Kadanoff, Physics 2, 263 (1966)

[11] K. G. Wilson, Phys, Rev. B4, 3174, 3184 (1971)

[12] E. Brown, J. W. Essam and C. M. Place, J. Phys. C8, 321 (1975).

[13] D. C. Rapaport, J. Phys. C5, 1830 (1972).

[14] D. Stauffer, Z. Phys. B22, 161, (1975).

[15] T. C. Lubensky, Phys. Rev. B15, 311 (1977).

[16] A. S. Skal and B. I. Shklovskii, Sov. Phys. Semicond. 8, 1029 (1975).

[17] P. G. de Gennes, J. Physique 37, L1 (1976).

[18] H. E. Stanley, R. J. Birgeneau, P. J. Reynolds and J. F. Nicoll, J. Phys. C9, L553, (1976).

[19] T. K. Bergstresser, J. Phys. C10, 3831 (1977).

[20] M. J. Stephen and G. S. Grest, Phys. Rev. Lett. 38, 567 (1977).

[21] D. J. Wallace and A. P. Young, Phys. Rev. B17, 2384 (1977).

[22] M. Barma, D. Kumar and R. B. Pandey, J. Phys. C12, 2625 (1977).

[23] D. Kumar, R. B. Pandey and M. Barma, Phys. Rev. B23, 2269, (1981).

[24] R. B. Stinchcombe; J. Phys. C12, 2625 (1979): *ibid.* C12, 4533 (1979).

[25] D. J. Breed, K. Gilijamse, J. W. E. Sterkenberg and A. R. Miedema, Physica 68, 303 (1973); J. App. Phys. 41, 1267 (1970).

[26] P.W. Kasteleyn and C.M. Fortuin, J. Phys. Soc. Japan Supp. 26, 11 (1969); Physica 57, 536 (1972).

[27] T.C. Lubensky, in *Ill-condensed Matter*, edited by R. Balian, R. Maynard and G. Toulouse (North Holland, Amsterdam, 1979). This is a very useful review on this subject.

[28] R. Bidaux J.P. Carton and G. Sarma, J. Phys. A9, L87 (1976).

[29] M. E. Fisher, J. Chem. Phys. 44, 616 (1966).

Chapter 5

Domains and Interfaces in Random Fields

Prabodh Shukla

We analyze the energetics of domains and interfaces in the presence of quenched random-fields, particularly at the lower critical dimension of the random-field Ising model. The relevance of this study to experiments is also discussed.

5.1 Introduction

Uniformly ordered ferromagnetic states tend to break up into domains due to a variety of reasons. The most familiar reason is the competition between short-range exchange forces, and long-range dipolar forces. The domain structure has a lower free energy than the uniformly ordered ferromagnetic state. Imry and Ma [1] argued that quenched random fields in a system may also cause a uniform ferromagnetic state to break into domains. The main result of Imry and Ma is that there is a lower critical dimensionality for a system below which an arbitrarily weak, quenched random-field with average value zero would destroy the uniform ferromagnetic order even at zero temperature. They concluded that the lower critical dimensionality is equal to two for spins of discrete symmetry (Ising spins), and four for spins of continuous symmetry (XY or Heisenberg spins). The argument of Imry and Ma is intuitively appealing, but nonrigorous. It remained unclear for several years if the results obtained from the argument were correct. The controversy was generated by the existence of another argument based on a field theoretic method that predicted the lower critical dimensionality of a pure (i.e. not disordered) system to be two dimensions lower than the lower critical dimensionality of corresponding random-field system. The lower critical dimensionality of pure Ising model is equal to one, and that of continuous spin model is equal to two. Thus the dimensional reduction argument predicted the lower critical dimensionality of random-field Ising model to be three, and that of random-field continuous spin model to be four. It took several years to realize the error in the dimensional reduction argument, and to resolve that the results obtained on the basis of the Imry-Ma argument are correct.

In section 5.2, we define a classical spin model of ferromagnetism incorporating random-fields, and recapitulate the arguments due to Imry and Ma for the existence of domains in the ground state of the model if the dimensionality of the system is lower than a critical dimension. Quenched random-fields are actually realized in several experiments. In the following, we illustrate this with a few widely studied examples [2,3]. We also mention briefly the difficulties involved in the experiments, and in the interpretation of experiments. It is due to these difficulties that the controversy between the domain wall theory and the dimensional reduction theory could not be resolved for several years based on experiments alone.

Although the domain argument gives the correct value of the lower critical dimension, it becomes inconclusive exactly at the lower critical dimension. It fails to clarify if Ising spins in two dimensions, and continuous spins in four dimensions can support true long range ferromagnetic order. In section 5.3, we discuss in detail the random-field Ising model in two dimensions focusing on the nature of domain walls and their roughness in this model. We conclude that there is no long range ferromagnetic order in the two dimensional random-field model. Our presentation is largely based on the work of Binder [4] supplemented with numerical simulation of the model. One might also wish to examine more closely continuous spin models in four dimensions because domain wall arguments are inconclusive here as well. However, we do not go into this analysis, and not much is published on this problem. The reason is perhaps historical. The conflict between the domain wall theory and the dimensional reduction theory occurred in the case of lower critical dimension of random Ising systems, and therefore the bulk of the study has been on the random-field Ising model.

5.2 Domains in Quenched Random Fields

We first describe the random-field model for classical spins. A classical spin is an n-component unit vector; it is called an Ising spin if n=1, XY spin if n=2, and Heisenberg spin if n=3. Higher values of n may also considered. The limit $n \to \infty$ is known as the spherical model. Mutually interacting spins of continuous symmetry ($n \geq 2$) do not have a gap between the ground state and low lying excited states, while the spins of discrete symmetry (n=1) do have a gap. Therefore the qualitative behavior of continuous spin models is different from that of the Ising model. For simplicity, we shall focus on the Ising model. The model is defined on a d-dimensional hypercubic lattice. Each site is labeled

5.2. Domains in Quenched Random Fields

by an integer i, and carries a classical spin S_i that takes the value $+1$ if the spin is pointing along the positive z-axis, and -1 if it is pointing along the negative z-axis. Each site also carries a quenched random magnetic field h_i. The set of fields $\{h_i\}$ are independent identically distributed random variables with a continuous probability distribution $\phi(h_i)$. There is a ferromagnetic interaction J between nearest neighbor spins. The Hamiltonian of the system is

$$H = -J \sum_{i,j} S_i S_j + \sum_i h_i S_i \tag{5.1}$$

In the absence of the random-field, the ground state of the system at zero temperature has all spins aligned parallel to each other. The ground state is doubly degenerate. The energy of the system when all spins are pointing up is the same as when they are all pointing down. To be specific, we choose the state with all spins pointing up i.e. $S_i = 1$ for each site i, and focus on a block of linear dimension L inside the infinite system. The volume of the block is L^d. The number of lattice sites and therefore the number of spins contained in the block is also of the order of L^d. The surface of the block is of the order of L^{d-1}. If the spins in the entire block were to flip down simultaneously, the bulk energy of the block would not be affected because the spins are still aligned parallel to each other. However, the block would now appear as a defect in the background of up spins. The energy of this defect will be the surface energy of the block. The surface energy is of the order of JL^{d-1}. In the absence of thermal energy or random-fields, there is no possibility that the block will flip down spontaneously in the fashion envisaged above. Now consider the effect of random-fields. The energy of spin S_i in the field h_i is equal to $h_i S_i$ or simply h_i since $S_i = 1$. Thus the random-field energy of the block is simply the sum of L^d random-fields inside the block. This sum is equally likely to be positive or negative, and is of the order of $\sigma L^{d/2}$, where σ is the root mean square deviation of the random fields from their average value zero. The surface energy of the block is always positive. If the random-field energy is negative and overwhelms the surface energy, the block will flip down spontaneously to lower its energy. Imry and Ma argue that if $d/2 > (d-1)$, i.e. if $d < 2$, then for any σ, there will be a characteristic length L^* such that the bulk energy gain will dominate over the surface energy loss for $L > L^*$. In other words, domains will occur spontaneously if $d < 2$. The characteristic length L^* is given by the equation,

$$\sigma [L^*]^{d/2} = J[L^*]^{d-1}, \text{ or } L^* = [J/\sigma]^{2/(2-d)} \tag{5.2}$$

As may be expected, the characteristic size of the domain L^* decreases with increasing disorder σ. Although we have considered the formation of a single defect in a uniform ferromagnetic state, numerous domains of linear size L^* will be formed in the system if $d < 2$, and the uniform ferromagnetic state will be completely destroyed. The density of domains is inversely related to their characteristic area. On the basis of arguments presented above, Imry and Ma concluded that the lower critical dimensionality of the random-field Ising model is two, i.e. the model would not support a uniform ferromagnetic state if the dimensionality of the system is lower than two. The argument is inconclusive in two dimensions.

Equation (5.1) is readily generalized to continuous spins. The Hamiltonian is given by,

$$H = -J \sum_{i,j} \vec{S}_i \cdot \vec{S}_j + \sum_i \vec{h}_i \cdot \vec{S}_i \qquad (5.3)$$

The spin $\vec{S}_i$, and the field $\vec{h}_i$ are now n-component vectors. In the absence of the random-field, the ground state of the system at zero temperature is infinitely degenerate. We choose one of the degenerate states, and consider the energetics of the formation of a reversed spin block of size L. The random-field energy of the block is calculated as before, but the surface energy gets modified. In the case of discrete spins the surface of the block is sharp. It cuts a pair of spins that are oriented opposite to each other. Ising spins can only point up or down. They cannot "tilt". The energy of a pair of Ising spins across the wall is equal to J compared with $-J$ in the absence of the wall. Continuous spins can tilt and the system utilizes this flexibility to lower its free energy. The change from up orientation outside the block to down orientation inside the block is achieved gradually by bending away from the z-axis over a distance that is known as the width of the domain wall. The advantage of bending away from the z-axis is that each pair of nearest neighbor spins in a domain wall is nearly parallel, and therefore it is an energetically favorable way of creating a defect. Suppose the angle between a pair of nearest neighbor spins in a domain wall is equal to θ. The energy of the pair with reference to the ground state without any defect is therefore equal to $-2J(1-\cos\theta)$, or $J\theta^2$ in the limit $\theta \to 0$. If the width of the domain wall is ω (i.e. there is a chain of ω bent spins between an up spin and a down spin), θ is equal to π/ω, and the energy cost of a wall of thickness ω is approximately equal to $\omega \, (\pi/\omega)^2$, or π^2/ω. This shows that the energy cost of a wall goes to zero as the thickness of the wall goes to infinity. In a real physical system, there is often an axis of easy magnetization. The domains are oriented up or down along the axis of easy magnetization. Bending of spins away from the easy axis costs what is known as the anisotropy energy. The anisotropy energy increases with the width of the wall, and therefore limits the width of the domain wall. In our simple model, we have not considered the anisotropy energy term. However, recognizing that domain walls have to be finite, we may make the plausible assumption that the domain walls are of similar size as the domains. In this case, the domain wall energy (per unit area of the wall) is approximately equal to $J\pi^2/L$. Noting that the area of the wall scales as L^{d-1}, we find that the domain wall energy scales as JL^{d-2}. This is the energy cost of creating the domain. The gain from random-fields scales as $\sigma L^{d/2}$ as before. Comparing the two terms we find the lower critical dimensionality for continuous spins is equal to four. The characteristic length L_c^* of domains in

5.3. Relevance to Experiments

the continuous spin model is given by the equation,

$$\sigma[L_c^*]^{d/2} = J[L_c^*]^{d-2}, \text{ or } L_c^* = [J/\sigma]^{2/(4-d)} \tag{5.4}$$

5.3 Relevance to Experiments

A random-field is not merely a theoretical artifact. Defects and imperfections in magnets have the effect of creating a quenched random field in the system. We illustrate this with a few examples [2, 3] that have played an important role in investigating the effects of random-fields experimentally. The connection with experiments was first made by a theoretical argument due to Fishman and Aharony [2] who showed that a weakly diluted anti-ferromagnet in a uniform external field acts like a ferromagnet in a random-field. Consider a one dimensional Ising anti-ferromagnet. Each unit cell has two spins, say S_1 and S_2 that are oriented opposite to each other. The order parameter (staggered magnetization per unit cell) is $S_1 - S_2$, and the total magnetization is $S_1 + S_2$. The system is exposed to a uniform external magnetic field h that couples to the total magnetization. In a randomly diluted system, one of the spins in a cell may be missing. Then the external field would couple to the remaining spin instead of the sum. The remaining spin may be written as an even or odd linear combination of the total and staggered magnetization. We may write $S_1 = [(S_1 + S_2) + (S_1 - S_2)]/2$, and $S_2 = [(S_1 + S_2) - (S_1 - S_2)]/2$. This means that in cells where one spin is missing, the external field couples to total magnetization as well as the staggered magnetization. The sign of the coupling to the staggered magnetization (the order parameter) is positive if S_2 is missing, and negative if S_1 is missing. The sign of the coupling between the order parameter and its conjugate field is random because either spin can be absent with equal probability. We may conclude from this that the critical behavior of a randomly diluted anti-ferromagnet in one dimension will be similar to that of a ferromagnet in a random-field. The argument is generalizable to higher dimensions. An anti-ferromagnet without dilution has two sub-lattices on which the spins are oriented opposite to each other. One of the sub-lattice is aligned with the applied field. In the presence of dilution, locally the sub-lattice with most spins tends to align with the applied field in competition with the global anti-ferromagnetic order in the absence of dilution. As in the one dimensional case, the applied field acts as an effective random-field coupling to the order parameter (staggered magnetization).

The effective random-field produced by the applied field is proportional to the applied field. The strength of the random-field is therefore easily controlled. One can do scaling studies with varying strengths of disorder in a system by simply tuning the applied field rather than making fresh samples with different degrees of dilution. This simplifying feature has a far reaching significance in investigating a theoretical model (RFIM) experimentally, and the interaction between theory and experiment has helped the field develop considerably, and clarified many fine points of the model [5].

If there is an experimental system that fits a theoretical model well, one may think that the predictions of the model may be tested easily by experiments. For example, the domain argument predicts the lower critical dimensionality of RFIM to be two, and the dimensional reduction argument predicts it to be three, so we may determine the correct value by doing an appropriate experiment. However, the interpretation of experiments is often not straightforward. Concentration gradients in the diluted sample tend to round off a transition and affect the measurements of critical behavior drastically. Further, the majority of experiments are performed on samples prepared in the following two ways; (i) cooling it in zero magnetic field, and (ii) cooling it in a magnetic field, and turning the field off at the end. Experiments on samples prepared in the two ways give different results. Three dimensional field cooled samples show no long range order and were first thought to show that the lower critical dimensionality of RFIM is three. But three dimensional samples cooled in zero field showed long range order. So the question of lower critical dimensionality could not be settled by experiments initially. It was only after several years of controversy that the experimental situation resolved itself in favor of the domain argument, i.e. no long range order in two dimensions, but long range order in three dimensions. The main point that was clarified by theoretical work is that the field cooled state is not an equilibrium state. It relaxes logarithmically slowly, and one should not expect to see an equilibrium ordered state in a field cooled sample over any reasonable experimental time scale.

Rb_2CoF_4 is a good two dimensional Ising anti-ferromagnet [6]. It consists of layers of magnetic ions with a single dominant intralayer exchange interaction, and an interlayer interaction which is smaller by several orders of magnitude. It is very anisotropic so that the spins can be well represented as Ising spins. The material can be magnetically diluted by introducing a small fraction of manganese ions in place of cobalt ions. Crystals of $Rb_2Co_xMn_{1-x}F_4$ are good examples of a two dimensional diluted anti-ferromagnet suitable for an experimental realization of the two dimensional random-field Ising model. In three dimensions, the most studied dilute anti-ferromagnet is $Fe_xZn_{1-x}F_2$ crystal. In the pure ferrous fluoride FeF_2 crystal [7], the ferrous ions (Fe^{++}) are situated approximately on a body centered tetragonal lattice. Each ferrous ion is surrounded by a distorted octahedron of flurine (F^-) ions. The predominant interactions are a large single-ion anisotropy, and an anti-ferromagnetic exchange between nearest neighbor ferrous ions. The magnetic moments of ferrous ions on the corners of the tetragonal cell are anti-parallel to the magnetic moments of Fe^{++} ions on the body centers. The large crystal field anisotropy persists as the magnetic spins are diluted with Zn. The diluted crystal remains an excellent Ising anti-ferromagnet for all ranges of magnetic concentration x. Furthermore crystals with excellent structural quality can be grown for all concentrations x with extremely small concentration variation $\delta x < 10^{-3}$. These attributes combine to make $Fe_xZn_{1-x}F_2$ the popular choice for experiments on diluted anti-ferromagnets, although experiments have been done on several other materials as well [3].

5.4 Roughness of Interfaces

The discussion in Section 5.2 was based on domains having flat interfaces with other domains. For example, we considered a cube of upturned spins in a uniform ferromagnetic background. The faces of the cube defined the interface in this case. We could have considered a sphere of upturned spins in which case the interface would have been a surface of constant curvature. In both cases, the arguments for the stability of domains presented earlier would hold good. The question arises if our conclusions would hold even in the presence of rough interfaces with fluctuating curvature. It is conceivable that the freedom of a wandering interface in a random background would make it energetically more favorable for the system to break up into domains. In other words, the roughness of the interface could raise the lower critical dimensionality of the system from the value given by the domain argument. In order to address this concern, several workers reformulated the problem in terms of the interface. In this approach we focus on a single interface in the system. It is convenient to use a continuum model of the interface. We visualize the interface as an elastic membrane. A flat interface is described by a vector $\vec{\rho}$ in the (d-1)-dimensional hyperplane transverse to the z-axis. Fluctuations in the interface are described by a displacement $z(\rho)$ away from the flat surface ($z = 0$). The displacement $z(\rho)$ is a scalar that can take positive or negative values. The random-field is denoted by a function $h(\rho, z)$ with the property,

$$\langle h(\rho_1, z_1) h(\rho_2, z_2) \rangle = \delta(\rho_1 - \rho_2) \delta((z_1 - z_2), \tag{5.5}$$

where δ denotes Dirac's delta function. The Hamiltonian of the interface is written as,

$$H = J \int d\vec{\rho} [1 + (\vec{\nabla}_\rho z)^2]^{1/2} - \int d\vec{\rho} \int dz'\, h(\rho, z')\, \text{sign}[z' - z(\rho)] \tag{5.6}$$

The first term is the total area of the interface multiplied by the exchange energy J; it gives the energy penalty for creating the interface. The second term is the random-field energy gained in roughening the interface on the assumption that the spins are up above the interface and down below it. The Hamiltonian can be simplified by adding to it a constant term equal to

$$\int d\vec{\rho} \int dz'\, h(\rho, z')\, \text{sign}(z')$$

One gets,

$$H = J \int d\vec{\rho} [1 + (\vec{\nabla}_\rho z)^2]^{1/2} + 2 \int d\vec{\rho} \int_0^{z(\rho)} dz'\, h(\rho, z') \tag{5.7}$$

On length scales larger than the bulk correlation length, the slope of the interface $|\nabla_\rho|$ should be small in comparison with unity if the domain argument is to be valid. In the following, we shall assume that $|\nabla_\rho|$ is small, and explore the condition for the validity of this assumption. We therefore linearize the fluctuations in the interface to first order in $|\nabla \rho|$. We also omit an infinite constant that does not depend on the fluctuations in the interface. Thus we write

$$H = \int d\vec{\rho} \left[\frac{1}{2} J |\vec{\nabla}_\rho z|^2 + V(\rho, z(\rho)) \right]$$

$$\text{where, } V(\rho, z(\rho)) = 2 \int_0^{z(\rho)} dz' h(\rho, z') \tag{5.8}$$

Often the potential energy term $V(\rho, z(\rho))$ is regarded as an independent random variable with mean value zero and variance given by,

$$\langle V(\rho_1, z_1) V(\rho_2, z_2) \rangle = \delta(\rho_1 - \rho_2) R(z_1 - z_2), \tag{5.9}$$

Here $R(z_1 - z_2)$ is a sort range function of its argument [8, 9].

We characterize the roughness of the interface over a length $|\rho_1 - \rho_2|$ in the mean interface plane by a quantity $\omega(|\rho_1 - \rho_2|)$ that is defined as follows:

$$\omega^2(|\rho_1 - \rho_2|) = \langle [z(\rho_1) - z(\rho_2)]^2 \rangle \tag{5.10}$$

Over large length scales, the roughness $\omega(|\rho_1 - \rho_2|)$ is characterized by an exponent ζ defined as follows:

$$\omega(|\rho_1 - \rho_2|) = |\rho_1 - \rho_2|^\zeta \tag{5.11}$$

If the exponent ζ is nonzero, but less than unity, the interface is called a self-affine interface.

We shall not go into the detailed analysis of the interface Hamiltonian, but be content to give rough scaling arguments to show that the interface approach to the problem of lower critical dimensionality yields the same result as the domain approach. We approximate $|\vec{\nabla}_\rho z(\rho)|$ by $\omega(L)/L$, i.e. the ratio of the height of the interface to its linear extension. In this rather crude approximation the energy of the interface scales as

$$\Delta E = J L^{d-1} \left[\frac{\omega}{L} \right]^2 - \sigma \sqrt{\omega L^{d-1}}$$

The first term gives the energy cost of creating a fluctuation in a flat interface and the second term the energy gain from the fluctuation (the sum of random-fields under the interface). The two terms should balance each other in equilibrium. Differentiating ΔE with respect to the height of the interface ω, we get the scaling relations

$$\omega(L) = \left[\frac{\sigma}{4J} \right]^{2/3} L^{\frac{5-d}{3}}, \text{and } |\vec{\nabla}_\rho z(\rho)| = \frac{\omega}{L} \sim L^{\frac{2-d}{3}}$$

The above relations predict that the interface roughness is negligible if d > 2. The domain theory presented in Section 5.2 assumes that the interfaces are are not rough. We have now shown that this assumption is justified if the dimensionality of the system is greater than two. Actually, the interface approach is not fundamentally different from the domain approach. Therefore the interface analysis presented above should be seen as a consistency check with the domain approach, rather than an independent proof of the results of the domain theory. The central result of both approaches is that the lower critical dimensionality of the random field Ising model is equal to two.

5.5 Absence of Order in 2d RFIM

Consider an L×L square of down spins in a plane of up spins. Our task is to investigate if this defect or a similar defect can appear spontaneously in the two dimensional random-field Ising model? The energy of the defect (referred to the uniform state with all spins up) is equal to

$$\Delta E = 8JL - 2\sigma L$$

The first term is the energy required to create a domain wall along the edges of the square. Each of the four edges of length L requires an energy 2 J L. The second term gives the possible gain in energy (with probability half) from the random-fields inside the square. There are L^2 sites inside the square. The sum of random-fields inside the square is therefore approximately equal to ± σL. The sum is equally likely to be positive or negative. Only if the sum is negative, the block gains energy by flipping down. However, if $\sigma < 4J$, the gain from random-fields is not sufficient to compensate the loss from domain walls. Therefore at first thought it appears that the long range ferromagnetic order will remain stable in the presence of small disorder ($\sigma < 4J$).

We show in the following that sufficient energy may be gained from random fields to compensate the loss due to domain walls if roughening of the domain walls is taken into account. Suppose the square of down spins is placed in the first quadrant of the xy plane such that the bottom left corner of the square coincides with the origin of the coordinate system. For simplicity, we focus on the roughening of just one edge of the square, say the bottom edge along the x-axis from x=0 to x=L. Spins above the edge are down, and those below it are up.

Following Binder [4], we cut the wall into smaller pieces along its length and allow each piece to move up or down without bending. The cuts along the length are made in an iterative, self-similar manner on smaller and smaller length scales. First the entire wall is allowed to move up or down till it gains maximum energy from the random-fields. Then, the wall in its new position is cut into n segments of equal length. Each segment is allowed to move up or down in order to lower the energy of the system. The process is repeated by cutting each segment into n equal pieces again, and relaxing each piece without allowing it to bend.

Suppose the entire length L of the horizontal wall moves vertically by a distance ω when relaxed. The area covered by the moving wall is equal to ωL, and therefore the gain from random-fields is equal to $-\sigma\sqrt{\omega L}$. The wall may move up or down with equal probability. We consider the case when the wall moves down. In this case the patch of down spins increases in area from an L×L square to a rectangle L×(L+ω) rectangle, and an extra vertical wall of length ω has to be created on each side of the moving wall. This costs an energy equal to 2 J ω on each side. The total energy change in relaxing the wall is equal to,

$$\Delta E = 4J\omega - \sigma\sqrt{\omega L} \tag{5.12}$$

Minimizing the above expression with respect to ω gives

$$\omega = \left(\frac{\sigma}{8J}\right)^2 L \tag{5.13}$$

Substituting from equation (5.13) into equation (5.12) gives,

$$\Delta E = -4J\left(\frac{\sigma}{8J}\right)^2 L \tag{5.14}$$

Next we divide the wall into n equal segments. Each segment of length L/n is relaxed independently. It moves a distance ω_1, and gains an energy for the system equal to ΔE_1 given by,

$$\omega_1 = \left(\frac{\sigma}{8J}\right)^2 \frac{L}{n}; \quad \Delta E_1 = -4J\left(\frac{\sigma}{8J}\right)^2 \frac{L}{n} \tag{5.15}$$

Dividing each segment of length L/n into further n equal segments, and relaxing each smaller segment independently yields,

$$\omega_2 = \left(\frac{\sigma}{8J}\right)^2 \frac{L}{n^2}; \quad \Delta E_2 = -4J\left(\frac{\sigma}{8J}\right)^2 \frac{L}{n^2} \tag{5.16}$$

The process is continued till the typical distance moved by the smallest segment is of the order of unity. Note that distances less than unity (in units of lattice constant) do not have any physical significance on a lattice. Suppose it takes kmax iterations to reach the end, i.e.

$$\omega_{kmax} = \left(\frac{\sigma}{8J}\right)^2 \frac{L}{n^{kmax}} = 1, \tag{5.17}$$

or,

$$kmax = \left\lceil \frac{2\log\left(\frac{\sigma}{8J}\right) + \log L}{\log n} \right\rceil \tag{5.18}$$

The energy gained in the entire process of relaxation is given by,

$$\Delta E = \sum_{k=0}^{k=kmax} n^k \Delta E_k = -4J\left(\frac{\sigma}{8J}\right)^2 L \times (kmax + 1) \tag{5.19}$$

5.5. Absence of Order in 2d RFIM

Or,

$$\Delta E = -4J\left(\frac{\sigma}{8J}\right)^2 L \times \left[\frac{2\log\left(\frac{\sigma}{8J}\right) + \log L + \log n}{\log n}\right] \quad (5.20)$$

In the limit $\sigma \to 0$, and $L \to \infty$, the dominant term becomes

$$\Delta E = -4J\left(\frac{\sigma}{8J}\right)^2 \left[\frac{L \times \log L}{\log n}\right] \quad (5.21)$$

Thus the energy gained from random-fields by roughening the domain scales as $L\log L$. Roughening of the flat domain wall also increases its length. However, the increase in the length of the domain wall due to roughening is marginal in cases where the domain wall is well defined. In order to see this, we estimate the new length of the staggered contour describing the interface:

$$L_{new} = L + \sum_{k=0}^{k=kmax} n^k |\omega_k| = L[1 + \left(\frac{\sigma}{8J}\right)^2 (kmax + 1)] \quad (5.22)$$

$$= L\left[1 + \left(\frac{\sigma}{8J}\right)^2 \cdot \left\{\frac{2\log\left(\frac{\sigma}{8J}\right) + \log L + \log n}{\log n}\right\}\right] \quad (5.23)$$

The first term is the leading term in the limit $L \to \infty$, $\sigma \to 0$. The maximum excursion of the rough interface away from its initial flat position is estimated as,

$$\omega_{max} = \sum_{k=0}^{k=kmax} |\omega_k| = \left(\frac{\sigma}{8J}\right)^2 L \sum_{k=0}^{k=kmax} n^{-k} \approx \left(\frac{\sigma}{8J}\right)^2 L \quad (5.24)$$

The average square deviation from the flat interface is given by,

$$\langle\omega^2(L)\rangle = \sum_{k=0}^{k=kmax} \omega_k^2 = \left(\frac{\sigma}{8J}\right)^4 L^2 \sum_{k=0}^{k=kmax} n^{-2k} \approx \left(\frac{\sigma}{8J}\right)^4 L^2 \quad (5.25)$$

The characteristic length scale L^* over which the energy gained by the rough interface becomes comparable to the cost of making the interface is given by the equation,

$$\Delta E \approx 2JL - 4J\left(\frac{\sigma}{8J}\right)^2 \left[\frac{L \times \log L}{\log n}\right] \approx 0$$

The coefficients of the two terms in the middle are not intended to be exact. The important point is that the first term (the cost of making a domain wall) is linear in L, and the second term (the grain from roughening the wall in random-fields) is of the order of $L\log L$. The characteristic length L^* is the minimum length beyond which ΔE can become negative, and therefore the system can spontaneously break up into domains. Approximately,

$$L^* \approx \left(\frac{\sigma}{J}\right)^2 L^* \times \log L^*; \text{ or } \log L^* \approx \left(\frac{J}{\sigma}\right)^2; \text{ or } L^* \approx \exp\left(\frac{J}{\sigma}\right)^2$$

It is easily checked that over length scales L^*, the total length of the staggered contour of the interface is of the order L^* because $(\sigma/J)^2 \log L^*$ is of the order unity. This proves an assumption made earlier that there is a length scale over which the cost of the domain wall scales as L, and the gain from random-fields as $L \log L$. The gain from random-fields dominates over the cost of making a domain wall for $L > L^*$, and the system forms domains spontaneously. Hence we conclude that the two dimensional random-field Ising model does not support long range ferromagnetic order in the limit of infinite system size.

A few remarks are in order:

1. The energy gained from random-fields depends on a parameter n that is an artifact of our method of estimation of the energy. However this dependence is rather weak because it arises via a $\log n$ factor. It is not expected to seriously affect the conclusions drawn above.

2. Excursion of the wall at the k-th step may take it into a region already crossed by the wall at the previous step. Thus ΔE_k may have a contribution that is already included in ΔE_{k-1}. In other words, there may be a problem of double counting in our procedure. Binder [4] argues that this problem is not serious.

3. Binder [4] also argues that the conclusions drawn above would hold even if all cuts of the flat interface at the k-th stage are not exactly equal to L/n^k, but are equal to this value on the average.

5.6 A Toy Problem

The analysis presented in the preceding section shows that the energy gained by an interface from roughening in a random background scales as L log L, where L is the length of the interface. This result is extracted after several simplifying assumptions and approximations. It is therefore desirable to check the theoretical prediction against numerical simulations of the model. The difficulty is that the numerical simulation of the minimum energy configuration of an interface in a random background is almost as difficult as an exact theoretical solution of the problem. We therefore adopt an algorithm that has the virtue of being efficient, and we hope that it captures the essence of the physical problem. Simulation of our toy problem on systems of moderate size appear to bear out the theoretical predictions of the previous section [10].

We consider a square lattice of size L×L, and an interface that joins the bottom left corner of the square to the top right corner. It is convenient to consider the bottom left corner as the origin of the coordinate system; the bottom edge of the square as the x-axis; and the left edge as the y-axis. The interface is modeled by the path of a directed walk along the nearest neighbor bonds on the lattice. The walk is always directed along the positive y-axis and the positive x-axis. The interface made by such a walk does not have any overhangs or loops. For simplicity, we assume that the spins above the interface

5.6. A Toy Problem

are down, and those below it are up. The spins reside in the middle of plaquettes made by joining nearest neighbor bonds around square unit cells. The spins are conveniently labeled by cartesian coordinates (i, j) where i denotes the i-th row from below, and j-th column from left. A spin in the plaquette (i, j) experiences a random-field $h(i, j)$.

Our object is to find an interface that divides the up spins from the down spins in such a way that the net interaction energy of the spins with the random-fields at their sites is minimum. We consider the spins in one column of the square at a time. Suppose the interface in the first column separates i spins below the interface that are up, from $L - i$ spins above the interface that are down $(0 \leq i \leq L)$. The magnetic energy of the first column can be written as,

$$E(i, 1) = \sum_{i'=0}^{i'=L} h(i', 1) S_{i',1} = \sum_{i'=0}^{i'=i} h(i', 1) - \sum_{i'=i+1}^{i'=L} h(i', 1) \quad (5.26)$$

A computer algorithm generates random-fields $h(i, 1)$; and computes $E(i, 1)$ for each row in the first column $(i = 1, 2, \ldots L; j = 1)$. The same process is repeated for the second column, i.e. we generate random fields $h(i, 2)$, and compute $E(i, 2)$ for each row in the second column. Now a naive approach may be to put the interface at a position in the first column that corresponds to the minimum energy in the set $\{E(i, 1)\}$, and put the interface in the second column that corresponds to the minimum energy in the set $\{E(i, 2)\}$ subject to the restriction that the interface in the second column be placed at the same level or above the interface in the first column. Our algorithm described below is an improvement over this naive approach. We compute a quantity $E_{min}(i, 2)$ for each row in the second column from $E(i, 2)$ and $\{E(i, 1)\}$ as follows:

$$E_{min}(i, 2) = Min(0 \leq j \leq i) \left[E(i, 2) + E(j, 1) \right] \quad (5.27)$$

This process is recursed to calculate $E_{min}(i, 3)$ from $E(i, 3)$ and $\{E_{min}(i, 2)\}$; and so on.

$$E_{min}(i, 3) = Min(0 \leq j \leq i) \left[E(i, 3) + E_{min}(j, 2) \right] \quad (5.28)$$

After L recursions of the process we get,

$$E_{min}(i, L) = Min(0 \leq j \leq i) \left[E(i, L) + E_{min}(j, L - 1) \right] \quad (5.29)$$

The quantity $E_{min}(L, L)$ gives the minimum energy required to place an interface diagonally across the square. Equivalently, $E_{min}(L, L)$ is the maximum energy that an interface whose length is of the order of L can gain by roughening itself in the background of quenched random-fields.

Fig. 5.1 shows the results of numerical simulations on systems of linear size varying from $L = 5$ to $L = 100$. $E_{min}(L, L)/L$ is plotted along the y-axis against $\log L$ on the x-axis. The figure shows graphs for three distinct distributions of the random-field: (i) random-fields taking two discrete values

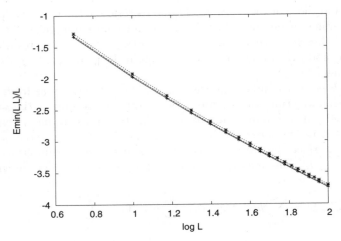

Figure 5.1: Energy gained by roughening of a diagonal interface in an $L \times L$ square. The energy gain divided by L is plotted along the y-axis against $\log L$ on the x-axis. Simulations for three different random-field distributions are shown. The lower curve is for a discrete distribution $h_i = \pm J$; the middle curve for a rectangular distribution of width $2J$ centered around the origin; the upper graph is for a Gaussian distribution of variance J. We have set $J = 1$, and scaled the middle curve by a factor $\sqrt{3}$ (see text). The figure suggests that the energy gain scales as $L \log L$.

± 1 with equal probability, (ii) a continuous rectangular distribution of width 2 centered around the origin, and (iii) a gaussian distribution of variance unity. The data has been averaged over 10^6 independent realizations of the random-field distribution. The figure suggests that the energy gained by the interface from random-fields scales as $L \log L$. The data for the rectangular distribution has been multiplied by $\sqrt{3}$. The modified data for the rectangular distribution (after multiplication by $\sqrt{3}$) lies very close to the data for the discrete ± 1 distribution. This may be expected because the standard square deviations of the two distributions are in the ratio $1 : \frac{1}{3}$.

5.7 Thermal Effects

At finite temperatures, interfaces are roughened by thermal energy as well as by quenched disorder. Let ξ_d be the length scale that describes roughening due to disorder at zero temperature, and ξ_T be the characteristic length over which the interface would roughen due to temperature if there were no disorder. As discussed in the previous section, ξ_d is the smallest portion of a straight line interface that moves sideways by a unit distance when the interface is relaxed

5.7. Thermal Effects

in a random background. We get from Eq. (5.17),

$$\xi_d = \left(\frac{8J}{\sigma}\right)^2 \tag{5.30}$$

The thermal length ξ_T is the typical distance one has to move along a straight interface before one finds a kink (i.e. a sideways turn of the interface) due to thermal fluctuation. The density of kinks is proportional to the Boltzmann factor $\exp(-2J/k_B T)$. Therefore,

$$\xi_T = \exp\left(\frac{2J}{k_B T}\right) \tag{5.31}$$

Let us consider a system with fixed disorder σ, and raise its temperature gradually above zero. At very low temperatures we have $\xi_T \gg \xi_d$ and therefore roughening of the interface is dominated by disorder. As may be expected, thermal fluctuations are essentially irrelevant at very low temperatures. At higher temperatures, when $\xi_T < \xi_d$, the thermal effects become relatively important and enhance roughening of the interface. Consider roughening by disorder at length scale L_k. The straight segment of length L_k will develop on average L_k/ξ_T kinks due to thermal fluctuations. Thus the mean square deviation $\langle \omega^2 \rangle_T$ due to thermal effects will be equal to

$$\langle \omega^2 \rangle_T = \frac{L_k}{\xi_T} = L_k \exp-\left(\frac{2J}{k_B T}\right) \tag{5.32}$$

The probability distribution of the interface width may be written as

$$P_T(\omega) \approx \exp\left[-\frac{\omega^2}{2\langle \omega^2 \rangle_T} + \frac{\sigma\sqrt{\omega L_k}}{k_B T}\right] \tag{5.33}$$

The first term on the right is the distribution corresponding to a completely random (Gaussian) walk. The second term takes into account the energy that can be gained by excursions in a random-field background. Minimizing $P_T(\omega)$ with respect to ω gives

$$\omega_l = \left(\frac{\sigma}{T}\right)^{\frac{2}{3}} L_k \exp\left(-\frac{4J}{k_B T}\right) \tag{5.34}$$

The energy gained from random-fields at length scale L_k is given by

$$U_k = -\sigma\sqrt{\omega L_k} = -\sigma\left(\frac{\sigma}{T}\right)^{\frac{1}{3}} L_k \exp\left(-\frac{2J}{3k_B T}\right) \tag{5.35}$$

Summing over contributions from all independent length scales, we get the total gain in energy U from roughening of a domain wall in a random medium at a finite temperature

$$U = -\sigma\sqrt{\omega L_k} = -\sigma\left(\frac{\sigma}{T}\right)^{\frac{1}{3}} L \log L \exp\left(-\frac{2J}{3k_B T}\right) \tag{5.36}$$

The leading term in the cost of making a domain wall still scales linearly with the length of the wall. At a characteristic length L^* the energy cost and energy gain terms balance each other. We get

$$L^* = \exp\left[\left(\frac{J}{\sigma}\right)\left(\frac{T}{\sigma}\right)^{\frac{1}{3}} \exp\left(\frac{2J}{3k_BT}\right)\right] \quad (5.37)$$

As noted earlier in this section, the above equation applies in the regime $\exp(J/k_BT) < (8J/\sigma)$. Random field systems with linear size greater than L^* will spontaneously breakup into domains at temperature T. Our discussion assumes that we have an interface with no overhangs and bubbles. This assumption is justified if the density of kinks is negligible, or equivalently if $\xi_T \gg 1$. If overhangs and bubbles are present in the system, it is not possible to formulate the problem in terms of a single-valued variable $z(\rho)$ describing the transverse fluctuations of a wall from a (d-1)-dimensional flat reference plane.

At higher temperatures, at around the critical temperature of the pure (i.e. without disorder) system overhangs and bubbles are no longer negligible. At these temperatures, a new length ξ_c comes into play that describes the diverging correlation length of the pure system at its critical point. We have,

$$\xi(t) = t^{-\nu}, t = \frac{T_c - T}{T_c}, \quad (5.38)$$

where T_c is the critical temperature of the pure system, t is the reduced temperature, and ν is a critical exponent. In a domain of linear size ξ_c, the net random field is of the order of $\sigma\xi^{d/2}$, but the spins within the domain are not aligned parallel to each other to the same degree as at low temperatures. The average magnetization per spin is of the order of t^β, where β is another critical exponent. Thus the gain from random-field energy is reduced to a quantity of the order of

$$U = \sigma\xi^{d/2}t^\beta = \sigma\xi^{\frac{d\nu-2\beta}{2\nu}} = \sigma\xi^{\frac{\gamma}{2\nu}} \quad (5.39)$$

Here γ is the critical exponent describing the diverging susceptibility at the critical temperature. We have made use of the well known relations, $\alpha+2\beta+\gamma = 2$, and $d\nu = 2 - \alpha$, where α is the critical exponent describing the singular part of the specific heat. Just as the thermal fluctuations reduce the gain from random-fields by a factor t^β, they also reduce rather drastically the cost of making a domain wall. The domain walls become ill-defined and spread out over a distance comparable to the linear size of the domain, and therefore the cost of making a wall is reduced from $J\xi$ to J only. The cost and gain become comparable to each other at a value of ξ given by

$$\xi = (J/\sigma)^{2\nu/\gamma} \quad (5.40)$$

The crossover to the length scale ξ takes place at a temperature given by the reduced temperature

$$t = (\sigma/J)^{2/\gamma} \quad (5.41)$$

5.7. Thermal Effects

In summary, as the temperature of the system is increased keeping the amount of random disorder σ fixed, we may see crossover phenomena between different length scales; a crossover from ξ_d to L^*, another from L^* to ξ, and finally from ξ to a distance of the order of unity at very high temperatures. It is plausible that the length scales are rather asymmetrical on the two sides of the transition temperature T_c of the pure system, and they remain finite and smaller than L^* in the vicinity of T_c due to slow relaxation of the system on experimental time scales. This speculation may possibly explain several observed phenomena in systems where the underlying lattice goes through a transition driven by a competition between temperature and quenched disorder.

References

[1] Y Imry and S K Ma, Phys Rev Lett 35, 1399 (1975).

[2] S Fishman and A Aharony, J Phys C 12, L729 (1979).

[3] D P Belanger, in *Spin Glasses and Random Fields*, ed. P Young, World Scientific.

[4] K Binder, Z Phys B 50, 343 (1983).

[5] T Nattermann, "Theory of the Random Field Ising Model", in *Spin Glasses and Random Fields*, ed. P Young, World Scientific.

[6] H Ikeda and M T Hutchings, J Phys C11, L529 (1978).

[7] M T Hutchings, B D Rainford, and H J Guggenheim, J Phys C3, 307 (1970).

[8] G Grinstein and S K Ma, Phys Rev Lett 49, 685 (1982), and Phys Rev B 28, 2588 (1983).

[9] D S Fisher, Phys Rev Lett 56, 1964 (1986).

[10] The idea of simulation was suggested by Prof D Dhar and carried out by Sumedha and the author during the course of these lectures.

Chapter 6

Vortex Glasses

G. Ravikumar

Irreversibility under field and temperature excursions and slow time relaxation of magnetization reveal the glassy behaviour of vortex matter in the mixed state of pinned or disordered type II superconductors. These phenomena are briefly summarized here.

6.1 Introduction

Superconductors exhibit zero resistance below a critical temperature T_c. However, the property that distinguishes them from perfect conductors is "reversible flux expulsion". Type-I superconductors expel flux completely from the interior of a superconductor when cooled below T_c in an applied magnetic field H smaller than a critical field H_c. This effect, known as the "Meissner effect", is characterized by the magnetization $M = -H$ [1–3] as shown in Fig. 6.1(a). It is empirically observed that the temperature dependence of H_c obeys $H_c(T) = H_c(0)[1-(T/T_c)^2]$. In the superconducting state, flux expulsion costs an additional free energy $\mu_0 H^2/2$ which compensates the condensation energy when the applied field equals the critical field H_c where superconductivity is destroyed. Condensation energy is the free energy difference $\mu_0 H_c^2/2$ between the normal and superconducting states (see reference [1]). Type-II superconductors exhibit a Meissner effect only below the first critical field H_{c1}. In the mixed state between H_{c1} and the second critical field H_{c2}, flux penetrates partially in the form of quantized vortices each carrying a quantum of flux $\phi_0 \approx 2.07\times10^{-15}$ Weber giving rise to the magnetization curve in Fig. 6.1(b). The resistance however remains zero upto H_{c2}. Vortices have a normal core of radius equal to the coherence length ξ where superconducting order parameter $\psi(r)$ is depressed as shown in Fig. 6.2(a). Around the normal core supercurrents circulate upto a distance equal to the London penetration depth λ. The field is maximum in the core and decreases over a length scale λ as shown in Fig. 6.2(b). Type I and

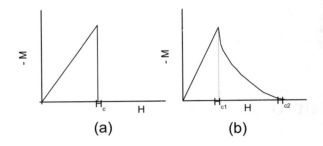

Figure 6.1: Magnetization M vs Field H curves of a type I (a) and type II (b) superconductors.

type II superconductors are characterized by the inequality $\lambda/\xi \leq 1/\sqrt{2}$ and $\lambda/\xi \geq 1/\sqrt{2}$ respectively. $H_{c1}(T) = (\phi_0/4\pi\lambda^2)\ln(\lambda/\xi)$ and $H_{c2}(T) = \phi_0/2\pi\xi^2$ constitute the magnetic phase diagram of a type II superconductor shown in Fig. 6.3. Thermodynamic critical field $H_c \approx (H_{c1}H_{c2})^{1/2}$ is the equivalent of H_c in type I superconductors. The temperature dependence of H_{c1} and H_{c2} is similar to that of H_c. The main focus of this article is the collective behavior of the interacting vortices in the mixed state $H_{c1}(T) < H < H_{c2}(T)$.

In the mixed phase, vortices interact by a repulsive interaction $U_{el} \approx (\phi_0/4\pi\lambda)^2 K_0(a_0/\lambda)$ ($K_0(x)$ is a modified Bessel function) to form into a triangular lattice with a lattice constant $a_0 \sim (\phi_0/B)^{1/2}$ where B is the flux density. As the field approaches H_{c2}, vortices overlap significantly. The magnetic state of an ideal type II superconductor, at a field H and temperature T is defined by its magnetization M which is independent of the path taken to reach a point (H, T) in the mixed state phase diagram [2]. It is completely governed by the interaction of the vortices. For instance, consider two different thermomagnetic histories. In the zero field cooled or ZFC, initially field is zero at a temperature $T > T_c$, i.e., the superconductor is in normal state. On cooling to $T < T_c$ and then raising the field to H, one obtains magnetization $M^{ZFC}(H,T)$. In the field

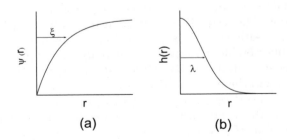

Figure 6.2: (left) Variation of order parameter $\psi(r)$ near a vortex. $\psi(r) = 0$ at the centre of the vortex and rises to unity over a length scale ξ. (right) Schematic field variation near the vortex.

6.1. Introduction

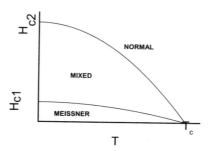

Figure 6.3: Phase diagram of a type II superconductor. $B = 0$ for $H < H_{c1}$ and the mixed state is between H_{c1} and H_{c2}.

cooled or FC case, field is applied at $T > T_c$ and the superconductor is cooled to T. The magnetization $M^{FC}(H,T)$ so obtained is equal to $M^{ZFC}(H,T)$ in an ideal type II superconductor.

However, such an ideal type II superconductor cannot support a bulk current and therefore not suitable for applications. A transport current of density $\mathbf{J}$ exerts a Lorentz force $\mathbf{F} = \mathbf{J} \times \mathbf{B}$ per unit volume of the vortex lattice. In the absence of any balancing force, vortices move in the direction of the Lorentz force, resulting in an electric field $\mathbf{E} = \mathbf{B} \times \mathbf{v}$ where $\mathbf{v}$ is the velocity of vortices. In the moving state, Lorentz force is balanced by a viscous drag exerted on the vortices by the underlying medium [2]. Fortunately there are various kinds of crystalline defects and impurities in superconducting materials which pin the vortices with a characteristic pinning energy $U_{pin} \approx \mu_0 H_c^2 \xi^3 / 2$ [2]. Thus there are now two energy scales in the mixed state, viz., U_{el} and U_{pin} mutually competing with each other. U_{el} favours an ordered arrangement of vortices into a two dimensional hexagonal lattice. On the other hand pinning tends to destroy this order.

Pinning arrests the vortex motion provided the current density J does not exceed a characteristic critical current density J_c, an important property from the application point of view. It marks the current above which dissipation sets in. A measurable voltage appears at a current I_c (see Fig. 6.4) in a Voltage (V) - Current (I) measurement and $J_c = I_c/A$ where A is the cross section of the superconductor perpendicular to current direction. Usually the $V - I$ measurements are performed in a configuration shown in the inset of Fig. 6.4, where the external field is applied perpendicular to the current direction, which is relevant to most high field applications of superconductivity. J_c macroscopically characterizes the distribution of the crystalline defects and their interaction with the vortices and the mutual interaction between the vortices in a vortex lattice. The development of various metallurgical processing techniques has made it possible to produce superconducting materials capable of carrying large non-dissipative currents, thus paving the way for generating high magnetic fields. However, there exists no theory which satisfactorily predicts J_c for a given distribution of pins.

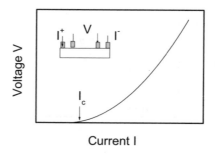

Figure 6.4: Typical voltage (V) - Current (I) characteristic of a superconductor. The arrow indicates I_c where a measurable voltage appears. Inset shows the configuration in which $V - I$ measurement is carried out.

6.2 Irreversibility in Type II Superconductors

In the absence of pinning, magnetization of a type II superconductor M is independent of the path taken to reach a point in the (H, T) phase space as already discussed in the last section. Free inward and outward mobility of the flux lines (vortices) readily enable the formation of the equilibrium state, which is solely determined by the mutual repulsion between vortices. Pinning restricts the free entry and exit of flux by trapping vortices in local free energy minima at the defect sites, thereby destroying the reversible behavior and it is no more possible to get a magnetization curve as in Fig. 6.5(b). The magnetization exhibits hysteresis under field [4] and temperature [5] excursions. A schematic magnetization hysteresis loop under isothermal field excursions is shown in Fig. 6.5 while in Fig. 6.6 typical magnetization under thermal cycling (FC and ZFC) is schematically shown with the external field held constant. From a microscopic view point, vortices arrange themselves into different metastable configurations depending on the field-temperature history. Because of the hysteresis in magnetization, pinned superconductors are called hard superconductors in analogy with hard ferromagnets. Details of the magnetic hysteresis under field and temperature excursions are reviewed in references [6], [7] and [8].

From the application point of view, a pinned superconductor is capable of carrying a dissipationless current (with zero electric field) of density J_c. While the current density J_c exerts a Lorentz force BJ_c per unit volume of the vortex lattice, the material defect structure is capable of resisting this force and prevent the motion of vortices ($E = 0$). Therefore BJ_c is called the pinning force density which is a charateristic of the material defect structure. For a current density larger than J_c, vortex motion causes dissipation. This is the basic mechanism behind the Voltage - Current characteristic shown in Fig. 6.4.

Magnetization hysteresis is understood by the phenomenological critical state model (CSM) proposed by Bean [4]. In this model, behaviour of the hard superconductor is somewhat similar to a perfect conductor with zero resistance.

6.2. Irreversibility in Type II Superconductors

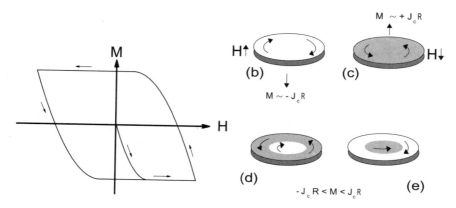

Figure 6.5: (a) Schematic magnetization hysteresis loop of a hard type II superconductor. Virgin magnetization curve (dotted line) meets the increasing field branch ($H \uparrow$) at $H = H^*$. $M \sim -J_c a$ on the increasing field branch ($H \uparrow$) and current distribution is shown in (b). Current distribution on the decreasing branch is shown in (c) corresponding to $M \sim J_c a$. Current distribution on branches B and D is shown in (d) and (e) respectively.

This model implicitly assumes that the resistance is zero for a current density smaller than J_c and infinite above. Any local variation of field induces a persistent current of density J_c in the superconductor which does not decay. The direction of the current induced is so as to oppose the field variation. Thus the only parameter in this model is the characteristic critical current density J_c of the material. In the original model, Bean assumed J_c to be field independent. But the field dependence can be easily incorporated into the model. More recent developments on the critical state model are reviewed in reference [8]. Let us now illustrate the model schematically. As shown in Fig. 6.5, a hard superconductor responds to an increasing field by an induced current in the clockwise direction which opposes the change in the local field. For a small applied field ΔH, current density is induced in an outer shell of thickness $\Delta H / J_c$

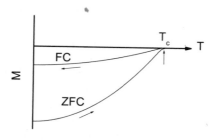

Figure 6.6: Schematic plots of $M^{ZFC}(T)$ and $M^{FC}(T)$. They are shown to be irreversible below T_c.

while the interior is completely shielded. As the field is raised further, field penetrates deeper and finally at an applied field $H^* \approx J_c a$ field and clockwise current of density J_c penetrate the entire sample. a is the sample dimension perpendicular to the field. This magnetization process of a virgin superconducting sample is shown by the dotted line in Fig. 6.5(a). Further increase in field does not alter the current distribution (J_c is assumed to be field independent). This results in a magnetization of magnitude $M \approx -J_c a$ (corresponding to field point A)as shown in Fig. 6.5(b). Let us now imagine that the applied field is raised to a maximum value $H_m >> H^*$ and then decreased. Magnetization then starts to increase (line B) as the current in an outer shell reverse from clockwise to anticlockwise direction (see Fig. 6.5(d). In other words, anticlockwise induced currents in the outer shell shield the interior where the clockwise currents are flowing initially. This process continues till the field is reduced to $H_m - 2H^*$ where current is reversed to anticlockwise direction throughout the sample. Further decrease of field does not affect the current distribution and the magnetization $M \approx J_c a$ (corresponding to field point C). Therefore at a given field the direction of the induced currents and the associated magnetization is dictated by the direction of field scan. This is the mechanism responsible for the magnetization hyseresis and the currents induced are persistent as the resistance is assumed to be zero for $J \leq J_c$. Magnetization on the increasing and decreasing branches of the hysteresis loop (corresponding to points A and C respectively in Fig. 6.5(a) are the limiting values that the magnetization of a hard superconductor can have. The intermediate magnetization values can be reached by changing the direction of field scan, for instance by traversing the curves B and D. Current distribution on the branch D is schematically depicted in Fig. 6.5(e).

An important result of this model is that the agnetization hysteresis due to these currents is roportional to J_c and the sample dimension a ransverse to the field. This basic result is extensively sed to estimate J_c from the experimental magnetization ysteresis loops using the relation

$$J_c \approx [M(H \downarrow) - M(H \uparrow)]/\mu_0 a \tag{6.1}$$

Here $M(H \uparrow)$ and $M(H \downarrow)$ are the agnetization in the increasing and decreasing field cans respectively. In the case of tiny single crystals, here it is difficult to make electrical contacts, J_c s inferred from magnetization hysteresis measurements (see reference [9] for further details).

Superconductors also exhibit hysteresis in magnetization M under thermal cycling. Consider a superconductor cooled in zero field (ZFC) to a temperature $T \leq T_c$. Application of a field H induces a magnetization $M^{ZFC}(T) \leq 0$, due to persistent current at a density $J_c(T)$. This state is far from equilibrium. M^{ZFC} decreases in magnitude as the superconductor is warmed up (see the discussion in references [6] and [8]) as a result of $J_c(T)$ decreasing with increasing temperature. On subsequent cooling of the superconductor in the presence of a field (field cooled or FC), no macroscopic currents are present in the superconductor and $M(T)$ follows the curve which is much smaller in magnitude compared to $M^{ZFC}(T)$ [6,7] as shown in Fig. 6.6. However, pinning still prevents the formation of equilibrium vortex configuration in the FC case as well. High T_c superconductors however exhibit reversible magnetization above a temperature T_{irr}. This would be discussed further in section 6.4.

6.3 Magnetic Relaxation and Flux Creep

While the inter-vortex interaction favours an ordered vortex lattice, pinning introduces spatial disorder in the vortex lattice. In conventional low T_c superconductors, U_{el} and U_{pin} are by far the most significant energy scales determining the physics of vortex arrays. In addition, thermal energy kT is responsible for temporal disorder in a vortex lattice and make the vortices fluctuate about their mean positions with the amplitude of fluctuation increasing with kT. By thermal activation, vortices overcome the pinning barriers and acquire mobility thereby reducing the effectiveness of pinning in arresting the vortex motion [10]. Important consequences are (i) electrical resistance [11] even for currents much smaller than J_c due to thermally activated vortex motion and (ii) slow logarithmic decay of persistent currents induced in the critical state, a phenomenon known as "flux creep" first proposed by Anderson [10,12]. Flux creep was first detected in hollow cylinders of conventional low T_c superconductors by Kim et al [13]. But in high T_c superconductors, the decay is dramatic (Giant Flux Creep) because of short coherence length (therefore smaller U_{pin}) and higher operating temperatures [14]. This is a major problem to be overcome before any high current applications of these materials are possible. The phenomenon of flux creep in high temperature superconductors was recently reviewed in reference [15].

Vortices interact through a long range repulsive interaction. It is therefore important to consider the collective behavior of the vortex lattice and its interaction with the pinning potential to describe the vortex dynamics. However, essentially a single particle picture [10,12] presented in the introductory text books is very illustrative in describing the thermally activated dynamics in a superconductor. In this picture, either single vortex segments or a bundle of vortices coherently but independently hop over a pinning barrier. Let us imagine a pinning potential as shown in Fig. 6.7(a) with an average barrier height U_0. In the absence of the Lorentz force ($J = 0$), thermal activation enables vortices to hop over the pinning barriers U_0 in all directions with equal probability resulting in zero net motion of vortices. However, in the presence of a Lorentz force due to a transport current $J \neq 0$, pinning potential is modified such that the pinning barrier is smaller in the direction of the Lorentz force and larger opposite to it (see Fig. 6.7(b). In other words, the barrier $U^-(J)$ in the direction of force decreases while the barrier $U^+(J)$ opposite to the direction of the force increases for increasing current density J. Unhindered vortex motion in the direction of Lorentz force commences at a current density J_c where the barrier $U^-(J)$ vanishes. But, vortex motion due to flux creep occurs even at

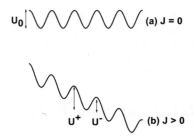

Figure 6.7: (a) Pinning landscape for $J = 0$ with average barrier height U_0. (b) Modified pinning landscape with $J \neq 0$ with different barrier heights along and opposite to the direction of force.

$J \leq J_c$. It smears the resistive transition or $V - I$ curves and makes the determination of J_c crucially dependent on the voltage criterion. The net vortex drift velocity in the direction of the Lorentz force is given by

$$v \approx w\nu_0 [e^{-U^-(J)/kT} - e^{-U^+(J)/kT}]$$

where w is the characteristic distance over which a flux bundle hops in a thermally activated event which occurs with a microscopic attempt frequency ν_0.

The first and second terms on the right hand side of the above equation are proportional to the probability rates of a vortex bundle hopping over the barrier in and opposite to the direction of Lorentz force respectively. In general $U^\pm(J)$ are non-linear functions of J. But, let us consider a simple case $U^\pm(J) = U_0(1 \pm J/J_c)$. Electric field E due to thermally activated vortex motion is induced in the current direction ($\mathbf{E} = \mathbf{B} \times \mathbf{v}$) and its magnitude is given by

$$E \approx Bw\nu_0 e^{-U_0/kT} Sinh[(U_0/kT)(J/J_c)] \tag{6.2}$$

The above equation clearly suggests that the resistivity E/J is non-zero and apparantly superconductvity is destroyed. Let us now discuss two limiting cases. First, in the regime of thermally activated flux flow (TAFF) corresponding to the limit $J \ll J_c$,

$$E \approx Bw\nu_0 \times e^{-U_0/kT}(U_0/kT)(J/J_c)$$

The temperature dependence in the above expression comes predominantly from the exponential term. The resistivity $\rho = E/J \sim e^{-U_0/kT}$ is non-zero even in the limit $J \to 0$. Palstra et al [16] measured linear resistance in the TAFF regime of $Bi_2Sr_2CaCu_2O_8$ crystal. Pinning energy U_0 is estimated from the slope of the straight line fit to the ρ vs $1/T$ data. They obtained the pinning energies to be few hundreds of Kelvin when the field is applied along the crystallographic c-axis and a few thousand Kelvin when the field is parallel to the ab-planes indicating that the pinning energies are much smaller than those in conventional superconductors. Thus, a measurement of temperature

6.3. Magnetic Relaxation and Flux Creep

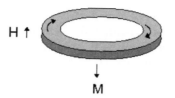

Figure 6.8: Induced currents in a superconducting ring of radius R, annular width ω and thickness d. Increasing the field induces clockwise currents.

dependent linear resistivity allows the determination of pinning energy barrier U_0, a phenomenological parameter depending on field and temperature. Microscopically, U_0 is a function of elastic properties of the vortex lattice and its interaction with the microscopic crystallographic defect structure.

The other important regime is when J is just below J_c. Here $U^-(J) \ll U^+(J)$ in which case magnitude of electric field E is given by

$$E \approx E_0 e^{(U_0/kT)(J/J_c)} \tag{6.3}$$

where $E_0 = Bw\nu_0 e^{-U_0/kT}$. We now show that this form of $E(J)$ results in a logarithmic decay of induced shielding currents [2,10,12,15], which can be best illustrated in a thin superconducting ring shown in Fig. 6.8.

Consider a thin superconducting ring of radius R and annular width ω and thickness d as shown in Fig. 6.8. Imagine a current of density J_c induced at some initial time $t = 0$ in the clockwise direction upon raising the external field to a value H. Within the critical state model, this current is persistent. But due to flux creep, there is an electric field $E(J)$ which increases exponentially with current as $E(J) = E_0 e^{\alpha J}$ ($\alpha = U_0/J_c kT$) which causes the induced current to decay. The decay is governed by

$$L(dI/dt) = -2\pi R E_0 e^{\alpha J} \tag{6.4}$$

where L is the inductance of the ring and $I(t) = J(t)\omega d$ and $J(t)$ is considered positive. Solving the above equation with the initial condition $J(t=0) = J_c$, we get

$$J(t) = J_c[1 - (kT/U_0)ln(1 + t/\tau)] \tag{6.5}$$

where $\tau = (kT/U_0)(LI_c/2\pi R E_0)e^{-U_0/kT}$. In general τ depends on the sample size and geometry, it's pinning strength and of course on the microscopic parameters ν_0 and w. For large time $t \gg \tau$, magnetization relaxation in superconducting specimens is in general described by

$$M(t) = M(t_0)[1 - R\ln(t/\tau)] \tag{6.6}$$

where R is the logarithmic decay rate proportional to kT/U_0, magnetic field and sample size [17].

An important question that naturally arises is whether there is a true zero resistance state in the presence of field. This is answered by vortex glass theory which suggests that the presence of weak point pins stabilizes an equilibrium vortex glass state at low temperatures where the barrier for vortex motion diverges in the limit of zero current [18] resulting in a truly zero resistance state. on the other hand at higher temperatures melting of the glass phase results in linear resistivity.

So far the flux creep phenomenon discussed above assumes either individual vortices or bundles of vortices coherently and independently hop over the pinning barriers in each thermally activated event. Fei'gelman et al [19] considered the collective interactions of the vortices subjected to weak point disorder which is important because of the long range nature of the intervortex interaction. They proposed various collective pinning regimes. They conclude that the effective pinning barrier depends on the current by the relation $U(J) = U_0[(J_c/J)^\mu - 1]$ where μ is an exponent depending on the field and temperature. Such an activation barrier, which again diverges for zero current, suggests that the resistance is zero in the limit of zero current(see reference [15] for a detailed discussion).

6.4 Irreversibility Line

As discussed earlier, $M - T$ plots in the FC and ZFC cases are usually irreversible below T_c (see Fig. 6.6). However, in high T_c superconductors, magnetization is observed to be reversible above an irreversibility temperature $T_{irr} \leq T_c$ as shown in Fig. 6.9. A plot of T_{irr} at different fields H is known as the "irreversibility line". Irreversibility line can also be obtained from $M - H$ loops measured at different temperatures T. The field where the magnetization hysteresis vanishes marks the irreversibility field $H_{irr}(T)$. Irreversibility line separates the mixed phase into reversible $(T \geq T_{irr}(H))$ and irreversible regimes $(T \leq T_{irr}(H))$. Reversible regime is characterized by $J_c = 0$ while $J_c \neq 0$ in the irreversible regime.

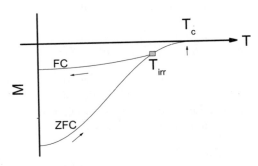

Figure 6.9: Schematic plot of M^{ZFC} and M^{FC} observed in high T_c superconductors. Above the irreversibility temperature T_{irr} magnetization is reversible.

6.4. Irreversibility Line

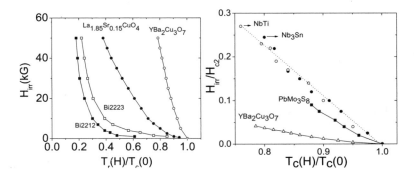

Figure 6.10: (a) Irreversibility lines $T_{irr} vs H$ for Bi-, La- and Y-based superconductors. (b) Irreversibility lines in various low temperature superconductors are compared with $YBa_2Cu_3O_7$. Data extracted from Ref. [20].

Yeshurun and Malozemoff [14, 15] recognized that the Giant flux creep is actually responsible for the large reversible region in high temperature superconductors. In this picture, induced currents in the ZFC state relax rapidly. However, above $T_{irr}(H)$, the relaxation is so rapid that the macroscopic shielding currents decay completely within the experimental time scales. Therefore, determination of the irreversibility line may crucially depend on the time scale of the measurement.

It was originally thought that the irreversibility line is a transition from solid to a liquid vortex phase. Such a belief is based on the knowledge that the solid phase has a finite shear modulus and therefore can be pinned by the underlying defect structure however weak. On the other hand, a vortex liquid can not be pinned due to lack of shear modulus. Current thinking is that irreversibility line only marks the depinning transition ($U_{pin} \approx kT$) In Bi- based compounds, irreversibility occurs only at much lower temperatures compared to the Y- based compounds although T_c is of the same order in both the systems(see Fig. 6.10(a)). Reversible region of high temperature superconductors is large primarily due to high operating temperatures and intrinsically weak pinning in these materials due to small coherence length (for a review on irreversibility lines in different low and high T_c superconducting systems, see reference [20]). Further, the role of thermal energy is significantly enhanced with increasing anisotropy. This is an important factor responsible for J_c being negligible over such a large part of the mixed phase diagram in Bi and Tl based cuprates [20].

In conventional low T_c superconductors, it is much harder to measure the reversible regime. Suenaga et al [20,21] have been able to detect a tiny reversible regime just below the $H_{c2}(T)$ line in very thin filaments of superconducting NbTi and Nb_3Sn [21], pointing to the fact that the role played by thermal energy is insignificant in conventional superconductors for most part of the mixed phase diagram. In Fig. 6.10(b), $T_{irr} vs H$ for conventional low T_c materials Nb_3Sn and $NbTi$ are compared with that of $YBa_2Cu_3O_7$.

References

[1] C. Kittel, Introduction to Solid State Physics, John Wiley and Sons, New York, USA 8th ed. 2000

[2] M. Tinkham, Introduction to Superconductivity, McGraw Hill, New York 2nd ed. 1996.

[3] P. G. de Gennes, Superconductivity in Metals and Alloys, W. A. Benjamin Inc., New York 1966.

[4] C. P. Bean, Phys. Rev. Lett. **8** (1962) 250; Rev. Mod. Phys. **36** (1964) 31.

[5] K. A. Muller, M. Takashige, J. G. Bednorz, Phys. Rev. Lett. **58** (1987) 1143.

[6] P. Chaddah, G. Ravikumar, A. K. Grover, C. Radhakrishnamurty and G. V. Subba Rao, Cryogenics **29**, (1989) 907.

[7] G. Ravikumar and P. Chaddah, Pramana-J. Phys. **31** (1988) L141.

[8] P. Chaddah, Pramana-J Phys **36** (1991) 353.

[9] P. Chaddah, K. V. Bhagwat and G. Ravikumar, Physica C **159** (1989) 570.

[10] P. W. Anderson, Phys. Rev. Lett. **9** (1962) 309; P. W. Anderson and Y. B. Kim, Rev. Mod. Phys. **36** (1964) 39.

[11] Y. B. Kim, C. F. Hempstead and A. R. Strnad, Rev. Mod. Phys. **36** (1964) 43.

[12] M. R. Beasley, R. Labusch and W. W. Webb, Phys. Rev. **181** (1969) 682.

[13] Y. B. Kim, C. F. Hempstead and A. R. Strnad, Phys. Rev. Lett. **9** (1962) 306; Phys. Rev. **129** (1963) 528.

[14] Y. Yeshurun and A. P. Malozemoff, Phys. Rev. Lett. **60** (1988) 2202.

[15] Y. Yeshurun, A. P. Malozemoff and A. Shaulov, Rev. Mod. Phys. **68** (1996) 911.

[16] T. T. M. Palstra, B. Batlogg, L. F. Schneemeyer and J. V. Waszczak, Phys. Rev. Lett. **61** (1988) 1662.

[17] P. Chaddah and G. Ravikumar, Phase Transitions, **19** (1989) 37.

[18] R. H. Koch, V. Foglietti, W. J. Gallagher, G. Koren, A. Gupta and M. P. A. Fisher, Phys. Rev. Lett. **63** (1989) 1511.

[19] M. V. Feigel'man and V. M. Vinokur, Phys. Rev. B **41** (1990) 8986; M. V. Feigel'man, V. B. Geshkenbein, A. I. larkin and V. M. Vinokur, Phys. Rev. Lett. **63** (1989) 2301.

[20] M. Suenaga, D. O. Welch and R. Budhani, Supercond. Sci. Technol. **5** (1992) S1.

[21] M. Suenaga, A. K. Ghosh, Y. Xu and D. O. Welch, Phys. Rev. Lett. **66** (1991) 1777.

Index 175

pinning energy, 161
Potts model, 129
pressure, 3
projection operator, 106
pure state, 41

quantized vortices, 159
quantum annealing, 49, 77
quantum fluctuations, 50
quantum phase transitions, 49
quenched disorder, 7

radius of gyration, 21
random exchange, 23
random field, 23
rare regions, 30
reaction field, 70
real space block renormalization, 59
recurrence relation, 62
relaxation phenomena, 93
relevant perturbation, 27
renormalization group, 18
renormalized spin variables, 62
replica, 36
replica method, 34
replica symmetry, 38
replica symmetry breaking, 41
RFIM, 145
RKKY interaction, 33

saddle point approximation, 13
saddle points, 104
satisfiability threshold, 46
satisfiable, 45
scale invariance, 18
scaling, 17
second critical field, 159
second law of thermodynamics, 3
self-averaging, 24
self-overlap, 42
Sherrington-Kirkpatrick Model, 34

simulated annealing, 77
site diluted magnets, 23
spherical model, 142
spin glasses, 30
spin waves, 55
stretched exponential, 94
strong glass, 90
supercooled luquid, 86
Suzuki-Trotter formalism, 49, 63

temperature, 3
thermal equilibrium, 1
thermal length, 155
thermodynamic critical field, 160
transport current, 161
transverse Ising model, 49
Transverse Ising spin glass, 68
trivial-non-trivial, 44
Trotter formula, 63
Type-II superconductors, 159
typical value, 25

universality, 18
universality classes, 18
upper critical dimension, 19

vector spin glasses, 68
VFT, 89
viscosity, 85
Vogel-Fulcher-Tammann-Hesse, 89
vortex bundle, 166
vortex core, 159
vortex glass, 168
vortex liquid, 169
vortex matter, 159
vortices, 159

words, 158

XY model, 23

zero-field cooled, 30
ZFC, 160

Fortuin, 22
Fortuin-Kastelyn mapping, 129
fragility, 89
free volume theory, 97
frustration, 25
frustration limited domain theory, 97
fundamental relation, 2

gap, 59
giant flux creep, 165
Gibbs free energy, 87
Gibbs state, 41
glass transition, 89
glassy state, 85
Griffiths, 29

H_c, 160
H_{c1}, 159
Harris Criterion, 27
Heisenberg equation of motion, 55
Heisenberg model, 23
Helmholtz free energy, 4
heterogeneous dynamics, 94
hyperscaling, 19

IMCT, 111
Imry-Ma argument, 28, 141
indirect exchange, 32
inherent structure, 100
interfacial energy, 87
internal constraint, 3
internal energy, 2
irreversibility temperature, 168
Ising spin models, 6

J_c, 161
Jordan-Wigner transformation, 73

K-SAT, 45
Kastelyn, 22
Kauzmann paradox, 92
KDP, 49
kinetic arrest, 97
kinetic facilitation, 111
Kob-Anderson model, 99
Kondo effect, 33
Kondo Hamiltonian, 32

KWW, 93

Landau-Ginzburg Theory, 13
Legendre transform, 4
Lennard-Jones potential, 95
Liouville operator, 106
literal, 45
London penetration depth, 159
lower critical dimension, 142

macrostate, 2
magnetic alloys, 30
magnetization, 159
magnetization hysteresis loop, 162
Mattis model, 25
mean cluster number, 122
mean field theory, 9
mechanical variable, 2
Meissner effect, 159
metabasin, 105
metallic glass, 86
microcanonical ensemble, 4
mixed state, 159
mode coupling theory, 93
molecular dynamics, 95
Monte Carlo simulations, 95
Mori-Zwanzig formalism, 105

Nishimori line, 68
node-link model, 127
NP-complete, 45
nucleation rate, 87

order parameter, 9

p-spin model, 46
pair-connectedness, 124
Parisi ansatz, 41
partition function, 7
Pauli spin matrices, 51
percolation, 20
percolation probability, 122
percolation threshold, 121
periodic boundary conditions, 95
phase space variables, 5
phase transitions, 9
pinning, 161

Index

α relaxation, 93

ac magnetic susceptibility, 30
Adam-Gibbs relation, 97
ageing, 94
Aharony, 145
amorphous solid, 86
Anderson Hamiltonian, 32
annealed disorder, 7
Arrhenius, 88

ballistic regime, 93
BCS, 57
Bean, 162
Bessel function, 160
Binder, 142
Boltzmann constant, 4
bond percolation, 20
Boolean variable, 45
boson peak, 93
Bragg-Williams approximation, 12

canonical ensemble, 6
cavity mean field theory, 34
chemical potential, 3
chemical variable, 2
clause, 45
cluster, 122
coherence length, 159
coherence volume, 27
computational complexity, 45
condensation energy, 159
connectivity length, 21
Constraint satisfaction, 45
cooperatively rearranging regions, 97
correlation length, 17
correlation matrix, 109

cost function, 77
critical cooling rate, 88
critical current density, 161
critical exponent, 18
critical nucleus, 87
critical state model, 162
crystalline nucleus, 87
cusp, 27

de Almeida-Thouless line, 44
Debye energy, 56
Debye relaxation, 93
dimension reduction, 142
dipolar, 33
droplet, 44
dynamical correlation length scale, 94
dynamical transition, 111

Edwards-Anderson model, 67
energy landscape approach, 97
ensemble average, 5
entropy, 2
entropy theory, 97
entropy vanishing, 101
equivalence of ensembles, 4
ergodicity, 5
excess enthalpy, 88
exchange interaction, 32
extensive variables, 2

Fermi wave vector, 33
field cooled, 30
first critical field, 159
Fishman, 145
fixed points, 62
flux creep, 165
flux expulsion, 159